D1433501

FUNDAMENTALS OF ENTOMOLOGY AND PLANT PATHOLOGY

SECOND EDITION

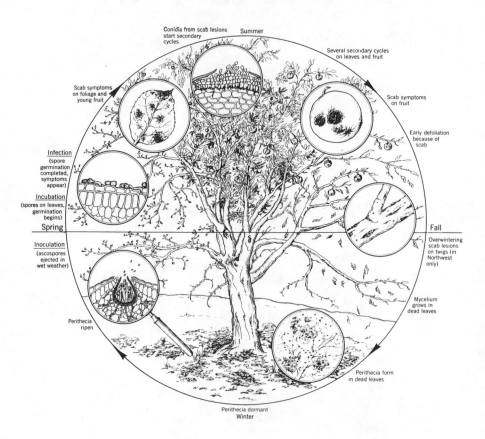

Conidia from scab lesions start secondary cycles

Summer

Several secondary cycles on leaves and fruit

Scab symptoms on foliage and young fruit

Scab symptoms on fruit

Infection (spore germination completed, symptoms appear)

Early defoliation because of scab

Incubation (spores on leaves, germination begins)

Spring

Fall

Overwintering scab lesions on twigs (in Northwest only)

Inoculation (ascospores ejected in wet weather)

Mycelium grows in dead leaves

Perithecia ripen

Perithecia form in dead leaves

Perithecia dormant
Winter

FUNDAMENTALS OF ENTOMOLOGY AND PLANT PATHOLOGY

SECOND EDITION

Louis L. Pyenson

Professor Emeritus,
State University,
Agricultural and
Technical College
Farmingdale, N.Y.

AVI PUBLISHING COMPANY, INC.
Westport, Connecticut

Library of Congress Cataloging in Publication Data

Pyenson, Louis, 1909–
  Fundamentals of entomology and plant pathology.

  Bibliography: p.
  Includes index.
  1. Insects, Injurious and beneficial.   2. Plant diseases.   3. Pest
control.   4. Pesticides.
I. Title.
SB601.P92   1980   632   79-16418
ISBN 0-87055-334-8

Preface to the Second Edition

I have been pleased with the general acceptance this book has found in the many colleges and universities in the country, and for the suggestions that have been made for improving this book.

As a result, this Edition has been enlarged to include descriptions and illustrations of all the major orders of insects. In addition the agriculturally and medically important families in these orders are described and illustrated to help students identify insects and appreciate their diversity. I have also included more detailed information on symphlans and on mite and tick families important to the health of livestock and humans.

I am greatly indebted to my colleagues, Dr. Harvey Barké, Dr. Gary Brown and several reviewers of the second edition for pointing out the areas where text presentation could be improved.

LOUIS PYENSON
Huntington, New York

November 1979

Preface to the First Edition

Many advances in the field of plant protection have taken place since this book was first published 26 years ago for the instruction of students interested in the agricultural and horticultural sciences. It included basic information on the nature and control of all organisms that harm crops—insects, plant pathogens, birds, rodents and weeds.

The present edition is limited to the study of insects, related forms, nematodes and plant pathogens. The book discusses the nature and control of these pests in as simple and concise a fashion as possible. Technical expressions and purely academic material have been kept to a minimum wherever possible; however, no science is simple and certain technical terms or expressions must be understood in order to comprehend information and developments in a given field. Agricultural production has become a highly technical profession and much more than manual dexterity and a strong back are required for success.

New knowledge, new pesticides, new concepts of pest control and greater concern with environmental preservation have greatly changed plant protection efforts in recent years. The need for a unified and integrated approach to pest control practices has become more apparent. Growers must now give greater consideration to environmental and public health impact of their practices. This book attempts to unify and integrate the knowledge in the fields covered. The last seven chapters of the book covering control practices and pesticide application are common to all phases of pest control.

The questions at the end of each chapter are presented to stimulate logical reasoning and discussion. The instructor may think of others. Answers to many of the questions may not be found in the text, nor are there any clear-cut answers in some cases, but the students should be encouraged to reason from the known facts. For more detailed information the references cited in the back of this book and experiment or extension service literature should be consulted.

The material is adaptable for an introductory one-semester course in plant protection (entomology and plant pathology). Laboratory studies ("Laboratory Manual for Entomology and Plant Pathology" by Pyenson and Barké, AVI Publ. Co. 1975) should supplement the textbook assignments. It is beyond the scope of this book to include specific information on many plant pests or troubles. Those included are mainly for the purpose of illustration. The subject matter lays the groundwork for a subsequent field-laboratory course on the identification and control of specific pests common to the students' own region. Although it is mainly intended for students, this book may also serve all those in agricultural pursuits as a helpful basic reference work.

I am greatly indebted to innumerable research workers and writers for the vast amount of information from which much of the material was gleaned. In the preparation of the manuscript I have profited from the suggestions and criticism of my colleague, Dr. Harvey Barké, and of several instructors in other colleges who used the original text in their own courses. Many of the excellent line drawings are by Mrs. Emily B. Steffens, a former student of mine. I wish to express my special appreciation to Mrs. Linda Galenskas for typing the manuscript and to my wife, Sheila, for her understanding and forbearance during my work on it.

It is also a pleasure to acknowledge the assistances given me by Mrs. Shirley A. DeLuca, AVI Publishing Company, in bringing out this book.

LOUIS L. PYENSON
Professor Emeritus,
Department of Biological
Sciences, State University,
Agricultural and Technical
College, Farmingdale, New York

January 1977

Contents

Prologue: The Problems of Plant Protection

The cultivated crop is the battleground where the interests of man meet and clash with those of innumerable organisms and phenomena of nature. Rodents may consume the plants, birds may eat their fruits, insects may drain their life juices, bacteria and fungi may invade their tissues, droughts may wither them, and weeds may crowd them out, often in spite of man's best efforts.

It is a battle of blind instincts and numbers against the inventive genius of man. Animals when hungry go in search of food, but man has learned how to grow plants for food for himself and his domestic animals. He has taken the most tasty, nourishing, or beautiful of these plants and learned how to reproduce, grow, and improve them, not in the manner of nature, in scattered units among thousands of other types of competing plants, but as millions upon millions of identical plants covering, in some cases, hundreds of acres. He has made it possible for a few to supply food for the many. This is agriculture, modern and intensive agriculture, which has enabled man to increase tremendously in population, to supply huge nonagricultural populations concentrated in large cities, and to advance his standard of living far beyond that of his cave man days.

But all this is contrary to nature's way of life, this hurrying and shaping plant evolution to suit man's own needs, this concentrating of single species of plants on the one hand, and of man and his domestic animals on the other. Nature works in centuries, but man in a period of 75 years has speeded up or by-passed natural evolution and has produced changes in North America that have led to the creation of many grave problems of plant protection. Our agricultural economy has reached such a stage that, if insect-, disease-, rodent-, and weed-control measures for our crops were suddenly stopped in the United States for a few years only, starvation would stalk the country, and huge segments of our population would be wiped out by the ensuing famine. Our highly developed agricultural and horticultural production has been made possible only by parallel developments in our knowledge of the various factors causing crop losses, and of ways of dealing with many of them quickly and effectively.

THE ECONOMIC IMPORTANCE OF PLANT PESTS

Food production is of primary importance to the ever increasing human populations of the world. Both insect pests and plant diseases limit this production. Approximately 80,000 insect species have been identified so far in North America. Some 10,000 of these 80,000 are known to be harmful to agriculture with 60 of the most important ones alone reducing crop yields annually by approximately 13% at a cost of 7.2 billion dollars in spite of pest control practices (Pimentel 1976). To these figures may be added some 10,000 infectious plant diseases with 500 leading ones of major crops reducing crop yield by approximately 12% at an annual cost of 6.6 billion dollars (Pimentel 1976). Noninfectious plant diseases caused by nutritional or environmental conditions undoubtedly add greatly to these losses. Post-harvest losses of stored foods to insects, microorganisms and rodents are estimated to be about 9% (Pimentel 1976). One should keep in mind, however, that these figures are annual averages and that losses tend to vary from year to year, from state to state, and even from field to field. Losses from insects and plant diseases in agriculturally undeveloped countries are generally much greater.

Even though the American growers use more than 1.2 billion lb of pesticides annually, controls for many pests are not adequate or effective and huge losses are still suffered. These figures present a challenge to growers to improve their methods of pest control.

THE RISE OF THE PLANT PROTECTION PROBLEM

This picture of costly crop losses due to pests is rather gloomy, but we must first understand how and why this problem has snowballed into its present state before we can comprehend the situation adequately.

Our fathers or grandfathers have often remarked that fine, clean crops were produced some 75 years ago without any need for sprays or spraying, and that plant pests did not trouble crops in those days. This is only partly true: many farmers even then had more than their share of insect and disease troubles.

Looking back at the records of pest troubles in the day of the horse and buggy we can find many records of destructive insect and disease outbreaks. In 1867, the Ohio Horticulture Society in its first report stated that many apples of that year's crop were defective and wormy and that the currant worm had invaded the state. During the 1870's, the plum curculio was mentioned many times as one of the worst pests of plums, cherries, apples, and peaches. Some growers even gave up plum growing because of this pest. The codling moth was first mentioned by name in 1868. In 1869, the Colorado potato beetle in its eastward march had invaded Evanston, Ill. Cankerworms, aphids, apple maggot, squash vine borer, rose chafer, and grape phylloxera were also noted in reports of that time. By 1893, the San Jose scale had reached Ohio, and its subsequent damage was so great that many growers believed that fruit growing was doomed. From 1892 to 1916, bark beetles, grape berry moth, woolly apple aphid, and other apple aphids became pests on crops.

The early history of disease appearance closely parallels that of insect troubles. As soon as a crop was fairly well established, disease promptly appeared. About 1850, apples, peaches, and thousands of acres of grapes were planted in Ohio. As soon as fruit production started, disease epidemics appeared. Mildew became so destructive that by 1869 thousands of acres of grapes in southern Ohio were being abandoned. The grape industry suffered great losses yearly from black rot until the grape industry moved north to the Lake Erie region. Apple scab was very serious, causing complete crop losses in many sections. Bitter rot was also prevalent. During the 1867–1885 period, crop after crop was lost and hundreds of orchards and vineyards were abandoned. Only the discovery of Bordeaux mixture as a fungicide, and its use beginning about 1885, made fruit growing possible in some areas.

THE CAUSES OF OUR PLANT PEST PROBLEMS

Nature's Methods Versus Agricultural Practices

Our modern type of agriculture has helped create these plant problems. Experience has shown us that the more highly developed and intensive agriculture becomes in any country, the more complicated are its problems, and the more plant troubles arise. This is not unexpected, since man through agricultural practices has been able to create great radical changes in the biological world in a relatively short period of evolutionary time. This in turn has upset nature's intricate system of checks, counterchecks, and succession of populations. In this system no organism, plant or animal, is allowed to gain complete domination indefinitely; it is sooner or later checked by overcrowding, lack of food, climatic factors, disease, or its natural enemies, and other organisms may gain a period of ascendency, only to be replaced in turn by still others. These periods of abundance and scarcity of animal populations may appear at regular or irregular intervals and are referred to as cycles.

Cycles in animals may run anywhere from a few to 50 or more years apart. As examples, field mice have 3–5 year cycles, squirrels 5–6 year cycles, tent caterpillars about 10 year cycles, and cankerworms about 5 year cycles.

Nature tends to scatter plant species, making survival difficult for their disease and animal pests; agriculture, on the other hand, tends to concentrate millions of a single species, bred usually for their nutritious value, in definite areas year after year. This is an open invitation for food hungry organisms to attack, destroy, and multiply. Acre upon acre of weedless corn, potatoes, or apples offer the optimum conditions for the rapid multiplication and spread of any organism that finds one of these crops a suitable host.

Increase in Old Pests

Many of our old pests have increased greatly in population. This has happened in spite of constantly improving methods of combating them. The New York State Experiment Station reported that codling moth populations reached an all-time peak in 1944 in the eastern United States. Unsprayed plots of apples in 1924 showed 10% worm attacks, whereas in 1944 they showed 30% worm attacks. Among the factors given as probably contributing to this build-up were: (1) the heavy plantings of apples tend to concentrate codling moth and cause rapid increase in numbers during favorable seasons; and (2) the popularity of the McIntosh has kept orchards planted largely to a variety most attractive to this pest. At the present time unsprayed apple orchards or potato fields under conditions favorable to their pests could be expected to have close to complete crop losses.

Increase in Formerly Obscure Pests

Many obscure insects, diseases, and animal pests, at one time restricted in area of spread by the presence of their wild weed hosts, have turned to or have adapted themselves to related cultivated crops and have spread wherever these crops are grown. Fire blight, once a disease of the native wild crab apple in the Hudson Valley, transferred to the imported apple and has become a serious pest on them and pears as well throughout the United States. The Colorado potato beetle, once a pest of wild solanaceous plants, and the Mexican bean beetle, once a pest of wild leguminous plants in the southwestern United States, transferred to their cultivated relatives, the potato and bean respectively, and are now pests in nearly all areas where these crops are grown.

Imported Pests

Many new pests, both animal and plant, have been brought in from other countries mainly with plant materials for food or cultivation. The

European corn borer arrived in a shipment of broom corn from Europe about 1909; the Japanese beetle in soil around the roots of ornamentals from the Orient about 1916; the Dutch elm disease on imported elm logs from Europe about 1930; and the chestnut blight on imported nursery stock from China about 1909. Both the English sparrow (1850) and the gypsy moth (1869) were intentionally imported for supposedly useful purposes before their destructiveness became apparent. Johnson grass was introduced as a forage plant from Turkey about 1830 and has since spread to most southern agricultural areas as a troublesome weed. These imported pests tend to flourish and spread rapidly if the climate is favorable, because most of their natural enemies are left behind.

New Strains and Races of Insects and Plant Diseases

In recent years, new strains or races of insects and plant disease-producing organisms have developed. This has lead to a constant changing of our plant protection problems. The continuous use of DDT and related compounds on houseflies has led to the development of resistant strains that retain this resistance through a number of generations. There is evidence that mites have developed strains resistant to organophosphates where they were used continuously. Continuous and frequent use of many of our newer organic pesticides has led to the appearance of resistant strains or races of insect pest populations or plant disease-producing organisms that can no longer be controlled by dosages formerly effective.

Through natural mutation and hybridization, new parasitic races of plant disease-producing organisms have originated, making plants that were formerly resistant or bred for resistance to a disease possible victims of the new strain. There are many records of supposedly resistant varieties being nearly wiped out by a disease epidemic or rendered useless for further crop production by a new physiological strain of the disease organism that has appeared on the scene. A new wheat variety, Ceres, was resistant to stem rust for 10 years, but became completely susceptible to it after 1935 because of the virulence of stem rust race 56 to it. More than 200 races of this disease are now known. Many disease-producing bacteria and viruses as well as fungi comprise races differing in virulence to different varieties of the same plant species. Thus we live under a constant menace that nature will create new combinations of genes for virulence to jeopardize plant varieties that now seem highly resistant under a wide range of conditions.

Solution of One Problem Often Creates Others

Plant varieties bred for resistance to one disease may often be very susceptible to others. Often major pests are created out of minor ones when new varieties of crop plants are grown. In the spring wheat area, there has been a constant change in disease problems. The wheat varieties Preston, Haynes Bluestem, and Glyndon Fife, grown 60 years ago, were resistant to scab, loose smut, and leaf rust but were susceptible to stem rust and stinking smut. When the stem rust-resistant Marquis variety replaced the older varieties, scab became a major problem; and when the farmers turned to durum wheats, root rots and ergot became problems. When the rust-resistant varieties Kota and Ceres replaced the durums, stinking smut, loose smut, and leaf rust again became troublesome. When Ceres was ruined by a new strain of stem rust in 1935, it was replaced by Thatcher, which again created an orange leaf rust and scab problem.

The apparent solution to the codling moth problem by the use of DDT sprays made major pests out of red mites and mealy bugs, which DDT did not control but whose parasites and predators it evidently did control.

Some newer varieties of plants may be more susceptible to certain pests or diseases unless bred specifically for resistance. Good examples are the old scab-resistant Baldwin apple and the newer scab-susceptible McIntosh apple.

Increase in Nutritional Disorders of Plants

There has apparently been a great increase in nutritional disorders of plants in recent years, or we have become more aware of them. This may be partly due to our agricultural practices of intensive cropping, lack of crop rotation, and soil exhaustion.

Wild Life Destroyed

With the occupation and elimination of wild areas, much of the beneficial wild life has been destroyed or driven away. Insectivorous birds, predatory birds, insectivorous animals, and animal predators are disappearing, leaving pests such as mice, rabbits, and insects which can live in close proximity to man to flourish and multiply.

Chemical Control of Insects Harmful to Natural Predators and Parasites

We have become more and more dependent on artificial chemical control to check our crop pests, and this in turn is harmful to their natural enemies, which under normal conditions help keep them in control. This destruction of beneficial populations made it necessary to continue chemical control once we committed ourselves to it to prevent huge losses.

Consumers Now Quality Conscious

Another grower's problem is that the consumers are now more quality conscious than ever before. Years ago no one had heard of graded fruit or vegetables—people expected to get the bad with the good—and a few cull or wormy apples were not considered just cause for turning down a possible purchase. Under present conditions, only the best grades get top prices and culls are discarded or sold at a loss. It does not pay the farmer to grow anything but the best grades of produce. Through economic necessity, growers have been made quality conscious, and they now must make special efforts to control pests to which they had formerly resigned themselves.

FACING THE PLANT PROTECTION PROBLEM

Whenever plants are grown, whether they are flowers or shrubs for beauty, trees for shade or lumber, or fruits, vegetables, or grains for food, a constant struggle must be waged to ensure the growth and production of these plants. The battle begins with the moment that seeds are placed in the soil and ends only with the cessation of the useful life of the plants or their products. No crop would do well if left to fend for itself for long.

To grow a crop successfully from beginning to end requires not only knowledge of the best cultural conditions but also awareness of all the potential dangers that may beset plants and ways to avoid the dangers. The successful farmer today must know what pests to expect on his crops, when and under what conditions their attacks will take place, the weak points in the pests' stages of development, the methods and materials best fitted to combat them, and finally the most effective and efficient equipment for applying control measures.

Individually and collectively we can no longer sit back in this new era of scientific development and expect to live by the pest control standards set by our fathers. To do outstanding jobs of pest control, growers must know how to use highly specialized information in order to put modern methods to work on their farms. The best farmers will be as scientifically trained and as well equipped with mechanical and chemical tools as most successful industrial leaders. Farming is rapidly passing from an art to a highly technical and mechanized operation.

Where to Obtain Information and Help

Quite often trouble appears with which the grower is not able to cope. He must then know where to get the services of agricultural scientists to help him solve his problem or problems. There are various agricultural agencies established to serve the grower's needs. Such agencies are county agricultural or farm bureau agents, state agricultural schools and colleges, state experiment stations, and the United States Department of Agriculture in Washington. Generally their own local, state or county agricultural agency will offer growers the quickest help and cooperation with their problems.

Keeping Up With Latest Developments

A part of every grower's spare time should be allocated to gathering information about his crops. This may be obtained through personal visits to agricultural institutions, attending lectures, and reading relevant extension literature. Every grower should be on the mailing lists of his local and Federal agricultural agencies and should subscribe to one or more agricultural periodicals to help him keep up with the latest developments, recommendations, and warnings.

It should be sobering to realize that man by his own efforts has never yet been able to eradicate one single insect species, plant disease, or weed from this earth.

QUESTIONS FOR THOUGHT AND DISCUSSION

1. Explain the statement that "agriculture is not natural."
2. Are plants helpless as far as fighting their own enemies is concerned?
3. How has man by-passed natural evolution?
4. The statement in the text, that starvation for us would result after only a few years if all pest-control measures were stopped, may sound fantastic. Is there any evidence to substantiate this statement?
5. Our highly developed state of agricultural and horticultural production was made possible only by parallel developments in pest control. Explain.
6. Why does increased agricultural production lead to more bugs and blights?
7. Under what conditions may it not be worthwhile to control the pests of a crop?
8. Would it be to our benefit to destroy all rodents and insects if we could?
9. Why do cultivated crops need care, whereas weeds flourish without care?
10. Why can we give only estimates of annual crop losses from pests?
11. Is a 3% average annual loss in a $100,000,000 crop as important as a 12% average annual loss in a $25,000,000 crop from an individual grower's standpoint? From a national standpoint?

12. Why is it misleading to say that crops were not troubled by pests in our grandfathers' time?
13. Some of our pests have increased greatly in numbers in spite of improved methods of control. Give some reasons why this has occurred.
14. Is it correct to say that the populations of organisms in nature are in a state of balance with each other?
15. What conditions would favor an obscure insect or plant disease in becoming a destructive plant pest?
16. Why is it generally true that an introduced pest is never as harmful in its native country as in its new habitat?
17. Is there any assurance that once an insect or a disease has been controlled or conquered by chemicals or by resistant plant varieties it will stay conquered? Present evidence.
18. Why have nutritional disorders of plants come to the forefront in recent years?
19. Is there any truth to the claim that natural organic fertilizers make plants resistant to pests whereas synthetic inorganic fertilizers tend to make plants susceptible to pests?
20. Does any relationship exist between wildlife abundance and pest abundance? Explain.
21. In what ways have the grading laws for fruits and vegetables helped in the promotion of pest control?
22. What are some of the uncertain factors in plant production?
23. Can you cite any examples of complete local or world eradication of an animal species? What made eradication possible in such cases?
24. Is it necessary for a grower to be better versed in plant troubles and their preventives now than in former years? Explain.

Chapter 1

The External and Internal Structure of Insects

Before we can deal intelligently and successfully with any plant pest, we must learn not only about its appearance but also something of its activities—the strong and weak points in its life cycle, especially its vulnerable periods—since during these our attacks will be most effective.

Insects show a remarkable range of adaptability to various environments. Some are aquatic, but most are terrestrial. They are found in every part of this earth where life can exist, from the arctic regions to the tropics. Their diet includes almost every kind of living and dead organic material from the tissues of plants to the blood of mammals. No part of a plant is safe from their attack, for they can riddle internally as well as suck and chew externally.

GENERAL BODY FEATURES OF INSECTS

All insects are covered by an outer skeleton (Fig. 1.1), a hard or soft chitinous cuticle, colorless and transparent in its pure state, as in the wings of a fly. Periodically, during their immature stages, insects wriggle out of their chitinous cuticles and form new ones, during what is known as a molt. The chitin, a nitrogenous polysaccharide, has properties similar to those of cellulose, the chief constituent of cotton and nylon.

This cuticle, in addition to supporting and protecting the internal organs of an insect, also prevents excessive loss of moisture from the body. The chitinous cuticle is resistant to the action of some of the strongest chemicals. This helps complicate the problem of insect control and may explain the ineffectiveness of many materials.

As aids in escaping their enemies, many insects have natural background-color patterns and spine-like or hair-like growths or protuberances. These natural colors may be acquired in three possible ways: the cuticle may be transparent and the color may be that of the blood or of the internal organs, as in many caterpillars; the cuticle may have colored

1

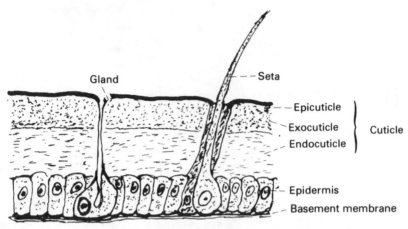

(Adapted from Wigglesworth and Snodgrass)

FIG. 1.1. DIAGRAMMATIC X-SECTION OF THE BODY WALL OF AN
INSECT (GREATLY ENLARGED)

scales or hairs growing on it, producing characteristic patterns as in most moths and butterflies; or the cuticle may be impregnated with pigments, as in many beetles.

Encased in this natural coat of armor, a hard-bodied insect still must have ample freedom of movement. This is possible because the insect's body is divided into a series of segments, one behind the other, with the cuticle remaining soft and pliable at each point of segmentation in the abdomen and at the joints in the legs and wings. Many insects, chiefly during their early development, have only a soft, somewhat elastic cuticle covering their bodies.

THE DIVISIONS OF THE BODY

The body of an insect as represented by the grasshopper (Fig. 1.2) and the beetle (Fig. 1.3) is divided into three well-defined regions—the *head*, the *thorax*, and the *abdomen*. The head is a sharply defined region bearing most of the sensory organs of the insect. The thorax is the heaviest portion of the body and bears the legs and the wings when present. The abdomen, which begins behind the point of attachment of the hind pair of legs, is generally the longest and most segmented part of an insect's body.

The Appendages of the Head

Every insect has a pair of *antennae* located on its head. These antennae vary greatly in length, shape, and number of segments in different species of insects (Fig. 1.4). They are sensory, but their exact function also tends to differ in different insect species. It is generally agreed that where they are long and flexible, as in grasshoppers, they function as

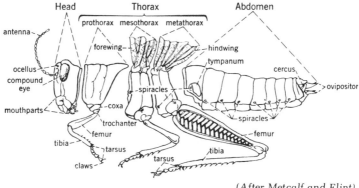

(After Metcalf and Flint)

FIG. 1.2. AN EXPLODED DRAWING OF A GRASSHOPPER SHOWING THE MAIN BODY REGIONS AND THEIR APPENDAGES

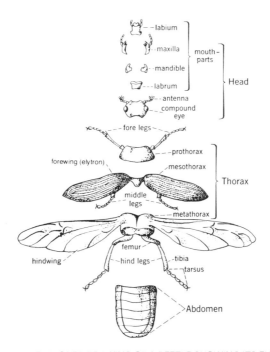

FIG. 1.3. AN EXPLODED DRAWING OF A BEETLE SHOWING ITS EXTERNAL STRUCTURES

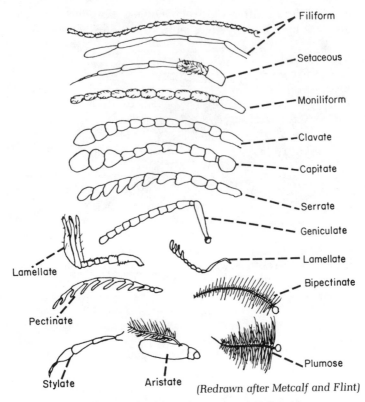

(Redrawn after Metcalf and Flint)

FIG. 1.4. MODIFICATIONS OF INSECT ANTENNAE

sensitive feelers for sounding out the environment. In flies, they have been found to have the sense of smell and, in male mosquitoes, they act as hearing organs vibrating to the pitch of the female mosquito. In other insects, they may be used to find food, locate mates, or, as in ants, for communication.

Nearly all adult insects and many immature ones have two *compound eyes* composed of hundreds of visual units called *facets*. Because the eyes bulge, insects can see in all directions at once without turning their heads or bodies. The extreme difficulty of approaching some adult insects from any direction without disturbing them indicates their excellent vision in all directions. These compound eyes vary greatly in size and shape in different insect species. In addition to the two compound eyes, most adult and immature insects have tiny, bubble-like visual units called *simple eyes* or *ocelli* on their heads. The usual

number of ocelli is 3 in adult insects and up to 6 pairs in some immature forms such as caterpillars which do not possess compound eyes.
The visual functions of ocelli in relation to that of the compound eyes
are not definitely known, but they are not considered organs of acute
vision and are reported to control light-response activities.

The remaining structures on the head are the mouth parts, which vary
greatly in structure and function. The structure of the mouth parts helps
tell us about the feeding habits of the insects, and these in turn are
closely related to control practices. Mouth parts are discussed in detail
later in this chapter.

The Thorax

Situated immediately behind the head is the thorax of an insect.
Typically, it is divided into three rather distinct segments, each segment
bearing a pair of legs (Figs. 1.2 and 1.3). The first thoracic segment, the
prothorax, never bears wings, but in most adult insects the second, or
mesothorax, and the third, or *metathorax*, each bears a pair of wings.
The three pairs of legs are generally alike in most insects but, in some,
one pair or another are specialized for a definite function, such as the
hind legs of the grasshopper for jumping, the front legs of a mole cricket
for digging and cutting roots, the hind legs of the honey bee for collecting pollen, and the front legs of the praying mantis for catching and
holding its prey.

The leg joints (Fig. 1.2) consist of the *coxa*, which fits into the body like
a ball and socket joint; the *trochanter*, a small, inconspicuous segment;
the *femur*, the heaviest portion of the leg; the *tibia*, the slimmest and
longest segment, sometimes armed with spines; and the *tarsus*, the foot
of the insect, made up of as many as five segments. To help in clinging,
the last tarsal segment generally bears a pair of claws and a pad-like
structure called the *pulvillus*, which in flies bears tiny, secretory hairs
with a mucilaginous substance to aid in clinging to ceilings and walls.

The Wings

Only adult insects have functional wings to carry them quickly from
place to place. Most mature insects have two pairs of wings, some have
only one pair, and a number are wingless throughout life. The wings
vary greatly in size, shape, and structure not only in different species but
also between the forewings and hindwings in the same species. The
wings are cuticular outgrowths from the body, bearing thickened, hollow areas called veins in characteristic patterns referred to as venation.
This venation remains constant for the same species and is of importance in identifying insects, as in some cases a slight difference in the

venation is apparently all that separates two very similar-looking groups of insects.

In many insects the forewings have lost their flying function and serve only as hard, protective covers, called *elytra* (sing. *elytron*) for the membranous hindwings (Fig. 1.3). In flight, the elytra are held out of the way of the hindwings, which fold underneath them when not in use. All beetles have elytra. The group of insects that are known as true bugs have forewings known as the *hemelytra*. As the name implies, their basal ends are thickened, whereas their distal ends are membranous. The wings of moths and butterflies are generally covered with scales which, when viewed under the microscope, look like shingles on a roof—arranged in rows and overlapping. True flies have only one pair of wings; the second pair is represented by a pair of slender, clubbed appendages called *halteres*.

The Abdomen

The segments that form the abdomen of insects are generally similar to each other but tend to taper near the end, where modifications occur in many forms.

In grasshoppers, hearing organs called the *tympana* are present on each side of the first lateral segment of the abdomen under the base of the hindwings. Male cicadas have a pair of "vocal" organs similarly located. Paired, small openings on the lateral sides of each of the abdominal segments except the last one or two are the openings of the breathing organs and are referred to as *spiracles*. One or two pairs are also found on some of the thoracic segments but never on the head. Thus we see that insects have many "noses," and even if some cease functioning the rest are ample to take care of the insects' needs. It is through these spiracles that many insecticides enter insects' bodies and kill them. Many insects have the power to open or close these spiracles, and this is thought to be one of the causes of the great variation in lethal effect of toxic gases on insects.

True legs are never present on an insect's abdomen but, in some immature forms such as caterpillars, paired, unsegmented appendages termed *prolegs*, used in clinging, may be present (Fig. 1.5). Many female insects have a structure on the distal end of the abdomen, known as an *ovipositor*, used for guiding and depositing eggs. Ovipositors are of various lengths, sizes, and shapes depending on the species of insect and on its manner of ovipositing. Insects that lay their eggs on and not in materials have no ovipositor, as would be expected. Some insects, notably the female wasps, bees, and hornets, have a sting at the end of the abdomen, mainly for defense but in some cases for paralyzing prey.

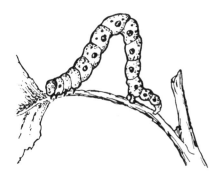

(Redrawn from Lutz, Field Book of Insects)

FIG. 1.5. A CATERPILLAR, SHOWING TWO PAIRS OF PROLEGS ON THE
ABDOMEN IN ADDITION TO THE NORMAL THREE PAIRS OF LEGS ON
THE THORAX

The Mouth Parts of Insects

The way insects eat and what they eat determines to a great extent the
type of control measures that may be effective against them.

The Chewing Type of Mouth Parts.—Insects' mouth parts show remarkable adaptations to their manner of feeding. The basic or primitive
kind is the chewing type which permits its possessor to bite off and chew
on, or into, external parts of a plant or to tunnel its way into some part of
the plant (Figs. 1.3, 1.6, and 1.7). Essentially it consists of a *labrum* or
upper lip which helps cover up the mouth parts from above; two heavy,
chitinized grinding organs, the *mandibles,* which work on a lateral
plane and help the insect tear off and crush its food; two slimmer,
sharper, chitinized organs, the *maxillae,* which work in unison with the
mandibles and help in tearing off bits of food; and one flap-like underlip,
the *labium,* which serves to close up the mouth from underneath. Each
maxilla has a fleshy lobe attached, which helps close up the sides of the
mouth and which bears a short, segmented, antenna-like structure, the
maxillary palp, with a sensory tip. The labium also bears a pair of shorter
palps, *labial palps,* with sensory tips. It is thought that these sensory tips
bear the organs of taste or smell, because the insects continuously
nuzzle their food with them while feeding. On the inner surface of the
labium is a tongue-like projection called the *hypopharynx* which is also
thought to have sensory functions. All insects with elytra and those that
in immature stages are "worm-like" have chewing-type mouth parts and
are usually controlled with a poison on their natural food that can take
effect when swallowed. Many of our most serious pests have this type of
mouth parts, with which they defoliate plants, bore into them, or cause
their fruit to be wormy.

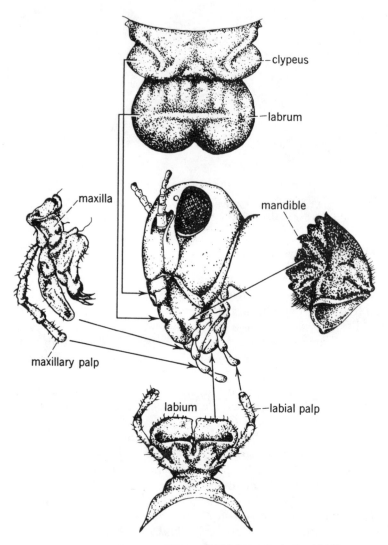

FIG. 1.6. THE CHEWING TYPE OF MOUTH PARTS AS TYPIFIED BY THE
GRASSHOPPER

The Piercing-Sucking Type of Mouth Parts.—The typical piercing-sucking type of mouth parts (Fig. 1.8) is thought to have evolved from the chewing type in the following manner: the labium gradually elongated and rolled up lengthwise to form a hollow tube, with a longitudinal slit opening which runs down its full front length like a gutter on a

(Redrawn from Britton, Conn. Agr. Expt. Sta. Bull. 344)
FIG. 1.7. A BEETLE CHEWING ON A LEAF

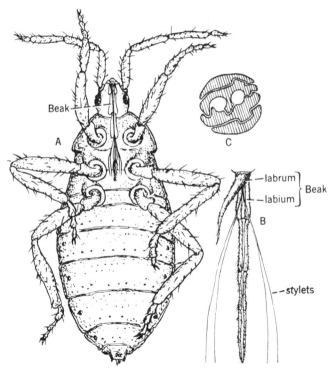

(Redrawn from Metcalf and Flint, Fundamentals of Insect Life)
FIG. 1.8. VENTRAL VIEW OF THE SQUASH BUG

A, Ventral view of the squash bug showing beak folded back against body; B, Beak enlarged, showing mouth parts; C, Cross section of stylets, showing how the two maxillae mesh together to form two capillary tubes (diagrammatic).

roof; the labrum also elongated but became a slim, V-shaped structure partly covering up the longitudinal slit on the labium. The two mandibles and the two maxillae elongated but became extremely hair-like and pointed at the tip, for piercing. These are given the name of *stylets* and are found inside the beak formed by the labium and the labrum. Both the maxillary and labial palps were lost in many species. The stylets fit closely together with the two maxillae on the inside and the two mandibles, slightly shorter, on the outside. The two maxillae are grooved on the inside in such a way that when they fit closely together they form two longitudinal capillary channels running into the head (Fig. 1.8C). One of these is for emitting saliva to mix with the plant juices before sucking them up, in much the same way that we mix our food with saliva for predigestion before swallowing it. The other is for sucking up the juice and saliva mixture into the stomach.

In plant-sucking insects the stylets alone pierce the wound; the beak acts only as a protective sheath and elbows up out of the way as the stylets sink in deeper (Fig. 1.9), leaving no visible hole when withdrawn. In some insect forms, the mouth parts or the saliva may carry the organisms of plant diseases or animal diseases, and these may be injected directly into the sap or the blood stream. Similar modifications of the chewing mouth parts produced such piercing-sucking mouth parts as are found in the blood-sucking flies, fleas, lice, and mosquitoes.

The piercing-sucking mouth parts are found in insects such as aphids, leafhoppers, scale insects, bugs, and cicadas. The injury to the plant from piercing-sucking insects results in discoloring and curling of foliage and weakening and drying up of twigs, limbs, and whole plants if the pests are abundant enough. Since these pests feed only from within, it can be readily seen that a poison deposited on the surface of their food and effective only when taken into the stomach would be of no value against them. Poisons that contact the insects' bodies are effective for these pests.

The Rasping-Sucking Type of Mouth Parts.—A group of tiny insects known as thrips have modified mouth parts that are referred to as rasping-sucking (Fig. 1.10). They have a short, conical beak with three stylets that move in and out, rasping the tissue so that juice exudes. This juice is then sucked up through the conical beak. Tissue attacked by these pests tends to take on a whitish, mottled appearance and may later look rusty. Because of this manner of feeding, thrips may be controlled by a poison placed on the plant surface and swallowed with the juices, and also by body contact poisons.

Sponging Type of Mouth Parts.—Another type of modified sucking mouth parts is found in certain flies, including the housefly (Fig. 1.11).

FIG. 1.9. A BUG FEEDING ON A LEAF

Their mouth parts consist of a hinged, fleshy proboscis which is partly concealed in a cavity under the head, with a sponge-like organ at the end for sopping up liquids. To distinguish it from other types of sucking mouth parts, it is referred to as a sponging type. It is interesting to know how these flies can feed on solid matter, as they often do, with such mouth parts: flies first let out saliva to dissolve or pre-digest the food,

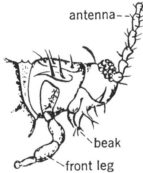

(*Redrawn from Metcalf and Flint, Fundamentals of Insect Life*)

FIG. 1.10. HEAD OF A THRIPS SHOWING CONICAL BEAK

and then they suck it up. This type of insect may be killed by poison taken internally or by poisons that come in contact with the body.

Sucking Type of Mouth Parts.—A simple sucking or siphoning type of mouth parts that is nothing more than a long tube is found in moths and butterflies (Fig. 1.12). This tube, also called tongue, is coiled like a watchspring in a groove on the ventral side of the head and is extended when the insect is feeding. Sometimes this tongue is much longer than the body of the insect; for example, hawk moths have a long proboscis

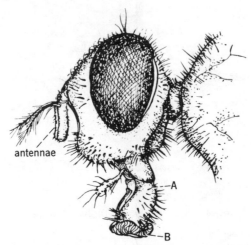

FIG. 1.11. HEAD OF A HOUSEFLY SHOWING A, THE FLESHY
PROBOSCIS WITH B, THE SPONGE-LIKE LABELLA

adapted for taking nectar from flowers with deep corollas. Moths and butterflies are not harmful in themselves, since they can feed only on liquids, but in many of their immature stages they are serious pests. Because of their manner of feeding and their food, control by stomach poisons of adult insects having these mouth parts is not practical. Control is directed mainly at the insects that have chewing mouth parts in their immature stages.

Chewing-Lapping Type of Mouth Parts.—Finally, there is a group of insects with mouth parts of a combination type known as chewing-lapping (Fig. 1.13). These mouth parts are found in many types of bees. The mandibles have retained their original function as chewing organs, but the maxillae and labium have been modified into a slim, lapping organ for taking up liquids, especially the nectar of flowers. Since these insects are mainly beneficial, control methods are not practiced on them but, on the contrary, controls used for others must be considered in the light of safety to many of these insects.

THE INTERNAL ORGANS OF INSECTS

Since all insecticides must take effect internally to destroy insects, it is important to know something of their internal make-up and of what may take place when certain insecticides are used.

Insects, though delicate looking, are remarkably hard to kill. In fact, in many instances they are much more durable than higher animals subjected to the same type of mistreatment. A knowledge of the internal

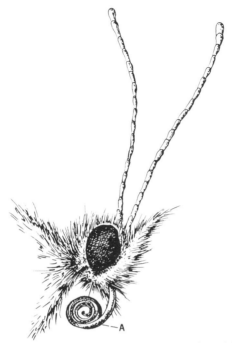

(Redrawn from Metcalf and Flint, Destructive and Useful Insects)

FIG. 1.12. HEAD OF A BUTTERFLY SHOWING A, THE LONG, COILED-UP
SUCKING TUBE (PROBOSCIS)

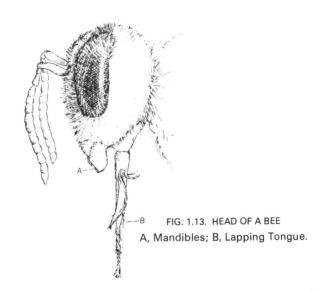

FIG. 1.13. HEAD OF A BEE
A, Mandibles; B, Lapping Tongue.

organs and of their function partly helps to explain why this is so (Fig. 1.14).

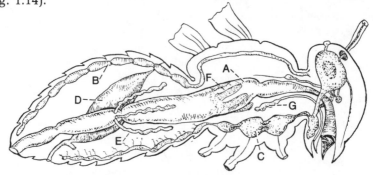

(Redrawn with modifications from Fernald, Applied Entomology)

FIG. 1.14. INTERNAL ORGANS OF AN INSECT (DIAGRAMMATIC):

A, Intestine; B, Heart; C, Central nervous system; D, Ovary; E, Malpighian tubules; F, Gastric caeca. G, Salivary glands.

Intestinal Tract.—The intestinal tract or alimentary canal in its simplest form is little more than a tube running the full length of the body (Fig. 1.14). Parts of this tube have become specialized for definite functions. In insects that feed on solid foods, the food particles are taken into the mouth where they are mixed with saliva and then passed on to the crop for storage. From the crop, the food passes into the proventriculus or gizzard, a muscular, chitinous organ for grinding up the food. From there the food passes into the large intestine, where some digestion takes place, and continues into the small intestine, which is generally made up of many coils, where the rest of the digestion takes place. The undigested material then passes into a temporary storage chamber, the rectum, and out of the insect's body through the anus. This type of intestinal tract is very similar to that of a bird and, like birds, insects excrete their urates in semi-solid form with the excreta. A stomach poison when it is swallowed by an insect normally takes effect in the large and small intestines. Some types of poisons may cause inflammation of the intestinal tract and slow death in a few days; others may be absorbed into the blood stream and may attack the nervous system, causing paralysis and quick death within a few hours.

Insects that feed on liquids only have no proventriculus and have a much shorter small intestine. Many insects such as bedbugs and carpet beetle larvae have the ability to withstand long periods of starvation and prolong their normal development up to a year or more.

The Circulatory System of Insects.—Insects have what is known as an open circulatory system as opposed to our closed one, where veins, arteries and capillaries conduct the blood. Insects have no veins or arteries; the blood merely circulates among the internal tissues from points of greater pressure to points of lower pressure. The heart is a simple pumping organ located along the mid-dorsal side of the insect; it consists of a series of chambers with valves and a tube extending into the head (Fig. 1.14). The valves suck in the blood from the abdominal cavity and by a series of contracting movements push it forward into the head, whence it gradually flows back into the abdomen because of the pressure differential. Accessory pulsating organs in the base of the legs of some insects help blood circulate into them. The heart can be seen in action in larvae with transparent chitin.

The blood of insects differs radically from ours in function. It does not serve like ours in respiration for carrying oxygen to cells and tissues and carrying carbon dioxide away, but it does carry foods from the intestines to the various tissues and cells. Since haemoglobin is generally absent in insect blood, it does not usually have the typical red color of our blood but is generally greenish or yellowish depending on the food dissolved in it. Insect blood has the power to clot, and therefore wounds are not necessarily fatal. The body temperature of insects is slightly above that of the environment and tends to parallel it—thus we speak of insects as being cold-blooded.

The Nervous System.—Insects have a very generalized type of nervous system that can stand much more abuse than ours (Fig. 1.14). It consists of masses of nervous tissue called ganglia, joined together by nerve cords. Each ganglion, through branching nerve fibers, governs the movements of the appendage and the sensations of the segment or segments in which it is found. The large ganglion in the head, which corresponds to our brain, is only a coordinating organ for all the other ganglia, so that movements are purposeful and not erratic. Injury to any one ganglion in the body will prevent movement and sensation only in that part which it serves and is not necessarily fatal. Insecticides that cause paralysis of insects evidently affect the whole nervous system and not just individual ganglia. Some insecticides cause a rapid paralysis of insects within a few minutes, whereas others may take as much as a week to become effective, with loss of coordination as the first symptom.

The Respiratory System.—Insects have no lungs or nostrils; instead they breathe through a series of paired holes (spiracles) along the sides of the body. Since insect blood does not conduct oxygen, a direct type of respiration is necessary (Fig. 1.15). This is accomplished by interconnecting tubes called *tracheae* that originate from the different spiracles and divide and subdivide into smaller and smaller tubes, the smallest of which are called *tracheoles*. These tracheoles ultimately reach every organ, tissue, and cell of the body, bringing them oxygen and carrying away carbon dioxide. The circulation of oxygen into, and carbon dioxide out of, the tracheae is mainly a matter of diffusion, with the exchange rate governed by the concentration gradient of the individual gases.

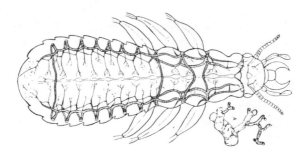

(Redrawn from Fernald, Applied Entomology)

FIG. 1.15. THE ARRANGEMENT OF THE MAIN TRACHEAL TUBES IN AN
INSECT (DIAGRAMMATIC) WITH AN ENLARGED PORTION OF THE
TRACHEAE

All spiracles are generally considered to have the same functions, and since they are interconnected by tracheae all but one or two may be clogged up and the insect will still be able to survive. Some insects may be kept under water for a day or more and may still recover when taken out, or they may be placed in a tightly closed bottle with food and live for days. Many contact poisons and all fumigants enter insects' bodies by way of the respiratory system. Since the respiratory system is directly connected to the nervous system, many of these materials cause quick paralysis. Insects may recover from this paralysis if they are not exposed to a high enough concentration or for a long enough time.

Reproductive Organs of Insects.—The reproductive organs of insects are located in the abdomen and, in the female, consist of a pair of ovaries with a common duct leading to an opening near the end of the abdomen (Fig. 1.14D) or into an ovipositor extending from the abdomen. The male reproductive organs consist of two spermaries and a duct leading to a chitinized tubular structure.

Organs of Secretion.—The organs of secretion in insects consist of the following types of glands adapted for various purposes:

The salivary glands are located in the thorax and open into the mouth parts. Because insect saliva sometimes carries disease-producing organisms, it is an important factor in their transmission to plants and animals. In many insect larvae, the salivary glands are modified to spin webs used in dropping down when disturbed, in swinging from tree to tree in a breeze, or in building a protective case (cocoon) before they enter the pupal stage. In the case of silkworms, their ability to spin silken threads is utilized commercially.

The scent glands are located in various regions of the bodies of some insects. They may produce distasteful or foul-smelling substances used in defense, as in stinkbugs, or attractive scents (pheromones) for luring the opposite sex, as in female moths.

Glands that produce waxy substances are found in many species, such as bees and scale insects. In scale insects, these glands may produce smooth, waxy coats covering the insect, or white, fluffy coats. These waxy, liquid-repelling coats make control of these pests very difficult, as they help protect the insects or their eggs from insecticides.

The defense glands used in connection with stings are found at the posterior end of the abdomen in insects such as bees, wasps, hornets, and some ants. The stings are sharp, hollow structures attached to a poison sac from which the poison is ejected into the wound. Some caterpillars, such as the saddleback caterpillar and the brown-tail moth caterpillar, have poisonous, nettling hairs, the venom in which irritates anyone brushing against them.

QUESTIONS FOR DISCUSSION
1. What are the advantages of an exoskeleton over an endoskeleton?
2. In what ways do insects differ from one another?
3. In what ways do insects resemble one another?
4. How do insects differ from the higher animals?
5. Are there any advantages in having six legs?
6. What parts of an insect's leg correspond to our own?
7. Of what significance is the distribution of the spiracles?
8. What are the structural features that are responsible for the success of the insect type of organization?
9. Name some external insect structures that show remarkable adaptations to their environment.
10. How could a knowledge of external structures and their functions help in insect pest control?
11. In general, what would be the difference between the manner of ovipositing in insects with ovipositors and those without?
12. State some sexual differences in insects.
13. How is it possible for some sucking insects to feed on solid food?
14. Why is it not advisable to spray fruit orchards during the blossoming period?
15. What are the differences between plant injury produced by chewing

insects and that by piercing-sucking insects? Which injures a plant more?

16. What type of mouth parts must a boring insect have? Why?
17. Some insects, like the clothes moth, do not feed in the adult stage. How do they survive and reproduce?
18. If an insect has a beak or snout, how could you tell whether it was piercing-sucking or chewing? Could you do so without looking at the mouth parts?
19. Can you give any reasons why curculios (snout beetles) are hard to control?
20. What use does a honey bee make of its double functioning mouth parts?
21. In what parts of plants is most of the food found? Does this correspond with points of greatest insect attack?
22. Why can a scale insect remain stationary throughout its life and still obtain ample food?
23. Why can an insect be kept alive for weeks in a tightly closed bottle?
24. What will happen if an insect's head is placed under water? If its body is placed under water?
25. What are the advantages of a generalized nervous system as found in insects?
26. Criticize the following statement: Insects react in certain ways because they wish to protect themselves or accomplish something.
27. Do you know of any instances when insects seemed to use the power of reasoning?
28. Name some methods utilized by insects to protect themselves from their enemies.
29. In killing insects, what are some of the ways in which we utilize their senses?
30. Name a problem in agriculture in which a knowledge of insect physiology would be valuable.

Chapter 2

The Development of Insects

There are more misconceptions about the stages of insect life and their relationships to one another than about those of any higher animal. This is because not all insects follow the same pattern of development as mammals. Each species has its own peculiar pattern of growth and development known as its metamorphosis.

Reproduction and The Egg Stage of Insects

Fertilization in Insects.—The life of each insect may be said to begin with the formation of an egg in the ovary of a female insect. As in the higher animals, fertilization of the egg by the sperm from the male insect is necessary in most species for the production of an individual. However, many female insects are capable of producing living, normal young without the necessity of mating. This evolutionary development, reproduction without fertilization, is referred to as parthenogenesis. As may be expected, some of our most important insect pests are parthenogenetic, with only females being produced. Males have never been found in many species such as the alfalfa snout beetle, the strawberry weevil, the white-fringed beetle, and some scale insects and thrips. In other species, notably aphids and some thrips, partial parthenogenesis occurs, some generations consisting only of females, with males appearing and fertilization taking place in occasional generations. In honey bees, the queen evidently has the ability to prevent fertilization of some eggs which develop into males or "drones," whereas the fertilized eggs produce workers (undeveloped females) or queens.

The Eggs of Insects.—The eggs of different species of insects are not as varied in size, shape, and appearance as the insects that lay them (Fig. 2.1). They are generally overlooked because they are very small, placed in concealment, and tend to blend with their surroundings. It is often very important to be able to tell the exact kind of insect that will develop from an egg, because many pests may be controlled in that stage, or controls may be prepared to take care of the susceptible stages of growth when they appear. The eggs of many species of insects can be recognized

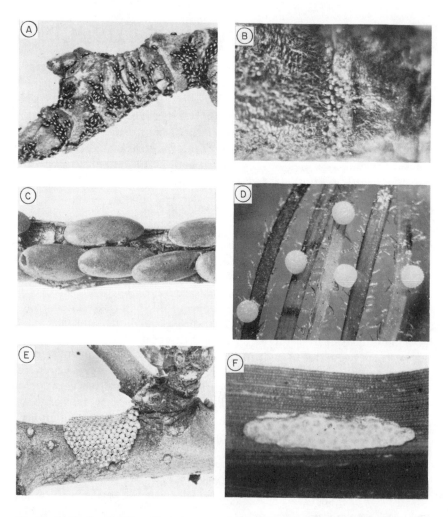

FIG. 2.1. EGGS OF INSECTS AND MITES: (A) APHID; (B) RED MITE; (C) KATYDID; (D) CORN
EARWORM (E) CANKERWORM; (F) CORN BORER (All greatly enlarged)

by careful examination and by consideration of distinguishing characteristics.

The egg of different species of insects differ in size, shape, color, location, attachment, kind of scar made in the laying, and sculpturing of the eggshell. In addition, some are laid singly and some in definite or indefinite masses; some are free and some are cemented together; some are uncovered and some are covered with hairs, cement, or paper-like, wax-like secretions.

Some insect eggs are so small that their characteristics can be seen only under a microscope, but all are generally laid in such a situation that the young when hatched may find suitable food. Sometimes, as in moths and butterflies, the adults deposit their eggs on food plants suitable for the larvae, although the adults themselves do not feed on these plants. There is no maternal instinct in insects except in a few exceptional cases where insects have been observed to guard their eggs, or with the social insects, such as bees and ants, which actually feed and take care of the young in the helpless, immature states of growth.

Eggs of insects may be laid all at one time, as by the tussock moths; a few a day for many days, as by the imported cabbage-worm; or a number of successive batches of eggs may be produced at intervals, as in the case of the squash bug and the Mexican bean beetle. The female insect generally dies shortly after she has completed egg laying.

Reproductive Powers of Insects.—Insects differ greatly in the number of eggs or young that a single female may produce. The number varies from a few dozen in some species to a million or more by the queens of social insects. The average number for most plant pests probably lies between 100–500, with many species laying as many as 1,000–2,000 eggs. This great reproductive potential far exceeds that of the higher animals, and many unbelievable examples of insects' reproductive powers have been cited under ideal conditions of food and survival.

One classic example is that of a pair of flies which could produce enough progeny during one active season (if they all lived) to cover the earth to a depth of 47 ft, allowing ⅛ cu in. to a fly. Still another example is that of a single aphid egg giving rise to 1,111,111,111 descendants in a single season, allowing 10 generations a season and only 10 young per aphid, all being parthenogenetic. The number of individuals that could theoretically be produced in a season, allowing 100 young per female and more than 10 generations a season, which is common in many species of aphids, "would be sufficient to cover the entire world completely with a continuous layer of aphids." Fortunately, natural checks prevent any such tremendous increases but, even with these natural checks which allow only a small percentage of individuals to die naturally, there are enough pests to cause great damage to crops and under favorable conditions in some seasons to cause tremendous losses.

Duration of The Egg Stage.—Under normal conditions, the time spent within the egg varies greatly with different insect species. It may last from only a few hours (as in certain flies) to 1–2 weeks or more (in most insects), depending mainly on temperatures. Unlike the eggs of birds, insect eggs do not require extra heat to begin development. Development begins when the temperature reaches a certain minimum (13–16°C) and increases in speed up to a definite maximum temperature (35–38°C), after which it tends to slow down. When temperatures drop below the minimum for development but above freezing for short periods, the eggs are not killed, and development continues when the temperature again rises. Thus Mexican bean beetle eggs may take 1 week to hatch when temperatures are about 29°C and 2–3 weeks or more when temperatures average around 18°C. Eggs of some insects such as tent caterpillars and aphids must undergo a freezing or a dry period before they will hatch. Many insects overwinter in the egg stage, all other stages dying off.

Many insects that lay their eggs exposed on trees and shrubs (aphids, scale insects, leaf roller, etc.) may be controlled in that stage by insecticides known as ovicides. For these pests, insecticides are generally applied in the dormant season when potent materials may be used without danger of injury to the plants.

Types of Insect Reproduction.—Three types of reproduction are found in insects: oviparous, ovoviviparous, and viviparous. Oviparous reproduction is reproduction by bringing forth eggs that, after various periods, hatch and produce young. The great majority of our insects reproduce in this fashion. In ovoviviparous reproduction, the eggs are retained within the body of the insect until they have hatched, and the insect gives birth to active young. Most aphids, except for one generation in the fall, and some scale insects, such as the San Jose scale, the soft brown scale, and the citrus mealy bug, are ovoviviparous. Viviparous reproduction, as found in man and in other mammals, is not common in insects. In this type of reproduction, the embryos develop in the body cavity and feed on certain secretions produced therein. When the embryos are fully mature, they are brought forth alive and active. The tsetse fly that causes African sleeping sickness and the sheep tick, a wingless fly that sucks the blood of sheep, are examples of this type of reproduction.

Metamorphosis of Insects

Metamorphosis refers to any change in form, structure, and appearance of an animal between birth and maturity. Insects may be divided into three groups with reference to a metamorphosis.

Insects Without a Metamorphosis (Ametabola).—These insects upon emerging from the eggs are essentially like the adult except for size. (Fig.

2.2). To become adult, they undergo a series of molts at intervals, each time increasing in size until the sexually mature stage is reached. Only two groups of insects have this type of metamorphosis: the Thysanura, which includes the household pests known as silver fish and fire brats; and the Collembola, which includes the springtails, one species of which attacks garden plants. These insects are wingless and are considered to be primitive types in the evolutionary scale of insects. They have chewing types of mouth parts.

Insects With a Gradual (Paurometabola) or Incomplete Metamorphosis (Hemimetabola).—All winged insects must undergo a metamorphosis during their development, since no insect has visible wings when it emerges from the eggshell. When insects with a gradual or incomplete metamorphosis hatch, they resemble adult insects but completely lack wings (Fig. 2.3). Except for some aquatic forms, they have the same types of appendages and structures and generally have the same habits and food plants as the parents. The young of the terrestrial (Paurometabola) forms are called *nymphs* and those of aquatic forms (Hemimetabola) are usually called *naiads*. Nymphs and naiads grow through a series of molts. Each time they molt they become larger. The stages between molts are called *instars*. The wing buds are scarcely

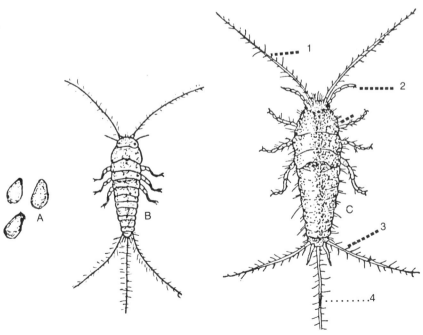

FIG. 2.2. NO METAMORPHOSIS: SILVER FISH (THYSANURA)

A, Eggs; B, Juvenile; C, Adult,-1, antenna; 2, maxillary palp; 3, cercus; 4, caudal filament.

noticeable on the very young, but after each molt they become larger until they expand into the fully developed wings of the adult after the final molt (Fig. 2.4). The number of molts and instars in insects undergoing a gradual or incomplete metamorphosis remains constant for the species but tends to vary greatly in different species. The time spent in each instar depends mainly on temperature and on nutritional conditions. Under normal summer conditions most insects with this metamorphosis take about a month to develop from egg to adult.

Most insects that have a gradual metamorphosis have piercing-sucking mouth parts, with the exception of the group known as the Orthoptera which includes the grasshoppers, crickets, and katydids. Insects having this type of metamorphosis include some of our worst pests, such as aphids, scale insects, leafhoppers, thrips, grasshoppers, and plant bugs. These insects are generally most easily controlled in the young nymphal stages; the adults tend to be much more resistant or may fly away from spray or dust applications.

Insects With a Complete or Complex Metamorphosis (Holometabola).—The young of insects with a complete metamorphosis

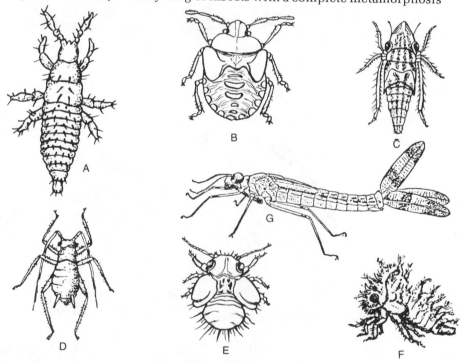

FIG. 2.3. NYMPHS

A, Thrips; B, Plant bug; C, Leafhopper; D, Aphid; E, Psyllid; F, Treehopper; G, Naiad of damsel-fly (note anal gills).

show no resemblance to the adult insects which they will become. They tend to be "worm-like" and are known as *larvae* in contrast to nymphs (Figs. 2.5, 2.6, and 2.7). Larvae show no traces of wings externally during any period of growth and may differ greatly in habits from the adult. Larvae also undergo molts and pass through several instars before they reach the size characteristic of their species. In structure they differ greatly from adults, although they have the general characteristics of insects. They tend to have elongated bodies that may bear accessory legs

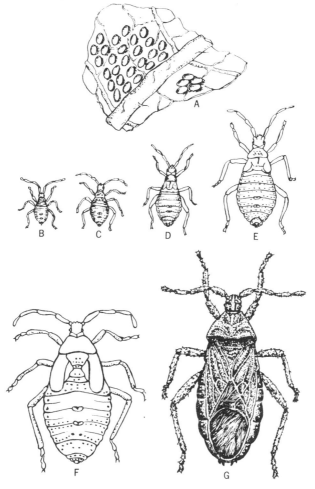

FIG. 2.4. GRADUAL METAMORPHOSIS AS ILLUSTRATED BY THE SQUASH BUG (ANASA TRISTIS)

A, Eggs on the underside of a leaf; B, C, D, E, and F, Five nymphal stages, the latter ones showing wing buds; G, The adult.

known as prolegs to help them in locomotion. They have only simple eyes; and the larvae of flies, bees, moths, and butterflies have mouth parts and feeding habits differing greatly from those of the adult. It is during the larval stages that these insects generally do the most damage, because that is when feeding and growth takes place.

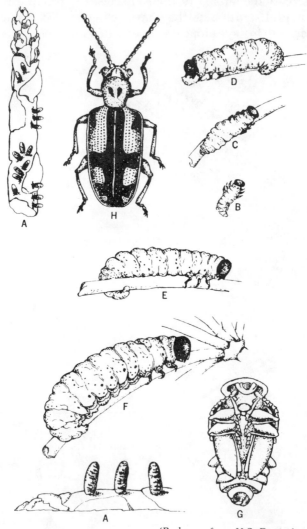

(Redrawn from U.S. Dept. Agr. Farmers' Bull. 1371, with modifications)

FIG. 2.5. COMPLETE METAMORPHOSIS AS ILLUSTRATED BY THE ASPARAGUS BEETLE *(CRIOCERIS ASPARAGI)*

A, Eggs on shoot; B, C, D, E, and F, Five larval stages; G, Pupa; H, Adult .

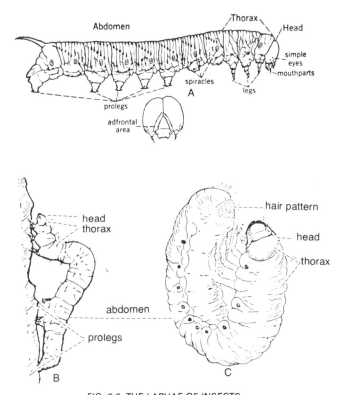

FIG. 2.6. THE LARVAE OF INSECTS
A, Hornworm or caterpillar (moth); B, Looper (moth); C, White grub (beetle).

Insect Pupae.—When a larva reaches its full growth, a striking change takes place. The larva sheds its skin, and all the body parts of the adult insect including wing buds appear folded up against the body. This is known as the *pupal stage,* and the insect is referred to as a *pupa* (Fig. 2.8). The pupa is quiescent and capable at the most only of wriggling its abdomen. It is during this stage that all the internal organs of the insect complete their change into the adult organs. When the change-over is complete the insect wriggles out of its pupal skin, expands its wings, and after a short period of drying is ready to fly and begin its adult life. The pupal stage of growth, because of its helplessness, is generally concealed in the soil or in some other protected place or bundled up in a silken case known as a *cocoon.* The fly larva uses its dry, hardened skin to form a pupal case known as a *puparium* (Fig. 2.8). The pupal stage is generally the most resistant to insecticides and also the least exposed, therefore control should be directed at the insects in more susceptible stages of growth.

Insect pupae are classified as: **obtect**—appendages held tightly glued in place; sometimes enclosed in a silky case or cocoon, e.g., moth pupae; **exarate**—appendages free but folded loosely against the body, e.g., many beetle pupae; and **coarctate**—pupae entirely enclosed in a cuticular case, a puparium, e.g., most flies.

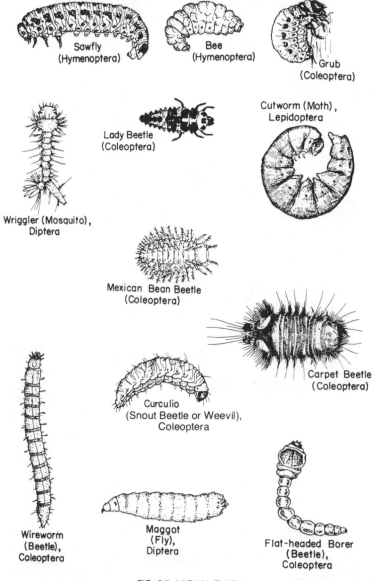

Sawfly
(Hymenoptera)

Bee
(Hymenoptera)

Grub
(Coleoptera)

Lady Beetle
(Coleoptera)

Cutworm (Moth),
Lepidoptera

Wriggler (Mosquito),
Diptera

Mexican Bean Beetle
(Coleoptera)

Carpet Beetle
(Coleoptera)

Curculio
(Snout Beetle or Weevil),
Coleoptera

Wireworm
(Beetle),
Coleoptera

Maggot
(Fly),
Diptera

Flat-headed Borer
(Beetle),
Coleoptera

FIG. 2.7. LARVAL TYPES

FIG. 2.8. THE PUPAL STAGE OF INSECTS

A, Moth; B, Beetle; C, Puparium of a fly; D, Cocoon of a moth.

Insect Growth.—Among insects, all increase in size takes place in the nymphal or larval stages that follow immediately after hatching. No fully mature insect, whether winged or wingless, increases in size. Once insects have acquired wings or are capable of reproduction they are considered adults. No immature insect has functional wings. This should make it clear that small beetles will not grow into large beetles, and that small flies will not grow into large flies. They are simply of a different species that characteristically attains that size, or sometimes they may be runts of the same species, caused by inadequate nutrition or unfavorable environmental conditions during growth.

Some types of larvae, because of their distinct characteristics, have been given definite names (Figs. 2.6 and 2.7). *Caterpillars* are the larvae of moths and butterflies; *grubs* are the larvae of beetles; *maggots* are the larvae of flies; *cutworms* are the larvae of a certain group of moths; and *wireworms* are the larvae of click beetles. Larvae have characteristics that make it possible to tell what type of insect they will develop into, and often the exact species can be recognized in its larval stage.

Identification of Immature Insects

A. Wing pads usually present, never functional, compound eyes present, sometimes brightly colored. Shape and appendages similar to adults. Legs usually large, two claws on tarsus. Mouth parts like parents'—chewing or piercing-sucking. *Nymphs and naiads.*
B. No signs of wings or wing pads, no compound eyes, body soft, thin-skinned, sometimes hairy, abdomen three or four times length of head and thorax together, shape "worm-like." Chewing type mouth parts, habits often very different from adults. *Larvae.*
C. Legs and wing pads encased in extra membrane and folded against

body, with some exceptions never used for locomotion. Inactive except for wriggling of abdomen by some forms when disturbed. Shape and appendages like adult. Often encased in cocoon, puparium, or soil. *Pupae.*

Identification of Agriculturally Important Larvae

A. With thoracic legs but without abdominal prolegs.
   1. Head distinct, sometimes depressed, mouth directed forward or downward, no adfrontal area. Larvae of *Coleoptera* (beetles).
   2. Head usually globular, adfrontal area usually present, mouth directed downward, spinneret present near end of labium. Larvae of *Lepidoptera* (moths, butterflies).

B. With thoracic legs and abdominal prolegs.
   1. Two, three, or five pairs of prolegs with tiny hooks (crochets), more than one pair of simple eyes or none on each side of head. Larvae of *Lepidoptera.*
   2. Prolegs, usually six to eight pairs, no tiny hooks, one simple eye or none on each side of head. Larvae of *Hymenoptera* (sawflies).

C. No thoracic legs, no prolegs, distinct head.
   1. Labium with a projecting median spinneret. Larvae of *Lepidoptera.*
   2. Body generally short, frequently U-shaped, pair of spiracles on each of the principle abdominal segments. Larvae of *Coleoptera* (curculios).

D. No thoracic legs, indistinct head, or head apparently wanting.
   1. Abdomen with several pairs of spiracles or none, no antennae, body tapers somewhat at both ends. Soft, white or yellow grubs found in wax, in paper cells, in bodies of insects, or in galls. Larvae of *Hymenoptera* (bees, wasps, ants).
   2. Abdomen usually with one pair of spiracles located on blunt end. Head end pointed. Mouth parts usually consisting of a pair of hooks. Larvae of *Diptera* (flies).

Pupae

A. With appendages usually fast to the body wall, often in cocoons. Abdominal segments often capable of much wriggling. *Lepidoptera.*
B. With appendages nearly always free, body wall generally thin, soft, and pale colored, rarely in cocoons. *Coleoptera.*
C. With appendages free, commonly encased in cocoons. *Hymenoptera.*
D. With appendages generally free, usually encased in a puparium. *Diptera.*

QUESTIONS FOR DISCUSSION

1. In what stages are insects usually harmless to crops? Why?
2. What insects are usually harmless to plants in the adult stage?
3. What characteristics of insect larvae usually determine whether they will develop into beetles, curculios, moths or butterflies, flies, or sawflies?
4. What larvae have no legs at all? How are they distinguished from one another?
5. Why is it important to know what a larva will become?
6. Give the main structural differences between a larva and an adult; a pupa and a larva; a nymph and an adult.
7. At what stages are insects usually most susceptible to insecticides?
8. In general, do nymphs and adults have the same habits? Do caterpillars and moths?
9. Why are insect eggs so inconspicuous?
10. Can an insect species be determined from eggs alone?
11. What is the value of recognizing insect eggs?
12. What may determine whether one generation of insects or several occur during the summer months?
13. Why is an insect considered an adult once it has acquired functional wings?
14. Is there any truth in the statement that little flies grow into big flies? Why?
15. At what stages are insects usually the most harmful to plants? Why?
16. Are all adult insects winged? Name some that are not.
17. If an organism is flying, and it isn't a bird or a bat, what is it?
18. During what stages does an insect grow larger and fatter?
19. What determines the number of molts a nymph or a larva will undergo before transforming to the next stage?
20. How does an insect grow?
21. What insects remain in one place during most of their life? What adaptations enable them to do this?
22. Differentiate between true worms and insect larvae.
23. At which stages of growth are insects the most defenseless? How are they usually protected in these stages?
24. Which type of reproduction in insects is more efficient: sexual or parthenogenetic?

Chapter 3

The Seasonal Life History of Insects

The seasonal life history of an insect is the record of all the stages through which the species passes, the approximate length of time spent in each stage, and the number of life cycles produced in a year. In addition, it is also a record of all that the insect does during its life. No grower can afford to overlook this information about his important crop pests if he is to deal with them efficiently and effectively.

The life history may generally be divided into an active and a dormant period (Fig. 3.1). The active period may in addition be divided into the

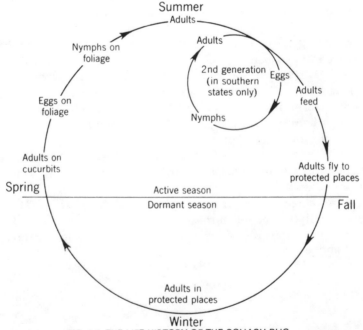

FIG. 3.1. THE LIFE HISTORY OF THE SQUASH BUG

primary life cycle or generation, beginning with the egg stage when activity starts in the spring, and successive similar secondary life cycles. In many pests there may be an overlapping of life cycles with all stages being present at the same time. This occurs mainly when adults reproduce over a long period or come out of hibernation over an extended period.

Insect Hibernation (Diapause)

The stage in which an insect overwinters varies among the different species of insects. Some insects completing the next-to-the-last as well as the last secondary cycle may produce the overwintering stage. For example, many codling moth larvae of the first brood in northern regions may not transfer into pupae but may remain dormant along with the second-brood larvae until the next spring. There are also cases where individuals of the overwintering stage may continue dormancy through the entire summer and finally complete their life cycle the following summer. The apple maggot which overwinters in a puparium in the soil has been observed to do this.

Insects generally hibernate in only one of their life stages, according to the species. They may hibernate in the egg stage on plants, in plant tissue, and in the soil; as full-grown larvae in protected places on plants, in plants, in the soil, and under objects; as adults in protected places in the woods, hedgerows, stone walls, and trash piles; or as partly grown nymphs in protected places. In tropical climates or in greenhouses, there may be no dormant period and generations may follow one another continuously. Where dry and rainy seasons occur, many insects go into a dormant stage (diapause) in the dry season similar to hibernation but referred to as estivation. The insects come out of estivation as soon as the rainy season begins.

Only those insects, such as aphids in the egg stage and scale insects, that hibernate externally on trees and shrubs may be controlled by sprays applied during the dormant season of the plants.

Insect Activities

The dormant insects become active when warm weather arrives in the spring—those that are immature continue their development, and those that overwinter as adults leave their hibernating quarters and seek food in fields and gardens. There they begin a new generation that goes through either a gradual or a complete metamorphosis according to their nature. The number of generations depends mainly on the inherent traits of the insects as well as on the weather conditions during the active season. Some pests may complete only one life cycle in a year (squash vine borer); some only one life cycle in 2 years (leopard moth); some only one life cycle in 3 years (common white grub); but many will complete as many generations in an active season as weather and climate allow. One of the longest-lived insects is the periodical cicada which takes as long as 17 years to complete its life cycle. Insects generally die shortly after they have lost their reproductive powers.

How Weather and Climate Affect Insects

Unfavorable weather is undoubtedly one of the main factors in leveling or reducing insect populations. Climate determines what insect pests are present and how many generations are possible in a single active season, but day-to-day weather plays a primary part in influencing insect abundance and damage.

The Effect of Temperature on Insects.—The general effect of temperature on insects is illustrated by Fig. 3.2. The optimum effective temperature range includes temperatures in which insects will thrive and multiply at the maximum rate. Optimum conditions vary with each insect species and even with different stages of the same species. Some insects such as the fall cankerworm adults and pear psylla adults are active and breed when temperatures are in the minimum effective range for most other insects; others such as thrips propagate best when temperatures

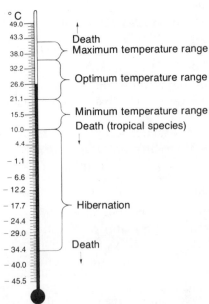

FIG. 3.2. THE RELATIONSHIP BETWEEN TEMPERATURE AND INSECT ACTIVITY IN TEMPERATE CLIMATES

are very high; but the majority of insects fall within the regions designated in Fig. 3.2.

The maximum effective temperature range includes temperatures at which insects will continue to live, but at which their development will be slowed down because of excessive heat; and the multiplication rate will be lowered because of increased mortality. If temperatures increase beyond this range, the insect will die. In the minimum effective temperature range, development is again slowed down because of cool conditions. Thus we may expect to have less trouble from most pests under very hot or very cool conditions, and a major amount of trouble when temperatures fall within the optimum effective range.

Cold Temperatures and Insects.—Tropical insects die if temperatures drop much below the minimum effective range, but insects of temperate climates go into a dormant stage known as hibernation. It is during hibernation that heavy mortality may occur from freezing or dessication. The ability of insects, like that of plants, to resist freezing is tied up with the undercooling point of the free- and bound-water content of their bodies. The exact subfreezing or undercooling temperature that insects in their hibernating stages may withstand varies with individuals and with different species, probably because of the variable ratio of free to bound water in their body tissues. Severe winters with extremely low temperatures and little snow cover on the ground and winters with many freezes and thaws induce high insect mortalities. Evenly cold winters with much snow greatly increase their survival. A number of insects, such as the corn earworm, that perish in the winter in the more northerly states overwinter successfully in the southern states and migrate north again in the spring and summer. The cotton leafworm does not have a dormant stage and migrates every summer into the southern cotton regions from points farther south in the tropics. The moths have been found as far north as New York State during the late summer.

Effect of Moisture on Insects.—In combination with temperature, moisture in both the air and soil also greatly influences the development of insects. A diagram similar to the one for temperature could be drawn indicating optimum, minimum, and maximum humidity ranges for insect groups or stages. Most insect species that attack plants cannot tolerate very low humidities or dry conditions. In the tropics insects go into estivation (diapause) in response to low humidities and high temperatures. Relative air humidities of from 50–80% are probably best for most external plant pests. The plants themselves probably help maintain favorable humidities for most plant pests attacking them internally. High humidities, together with normal or high temperatures, tend to induce bacterial or fungous disease epidemics among insects. Severe droughts are just as hard on some insect pests as they are on plants. Precipitation in the form of heavy, wind-driven rains is very destructive to small, delicate insects but, in the spring, precipitation in the form of warm rains helps bring insects out of hibernation.

Effects of Sunlight on Insects.—Both the intensity and the duration of sunlight also greatly influence insects. Some insects are active only at night, others are active only during the day. Some insects avoid bright light, although others favor it. The length of daylight has a profound effect on some insect species. Aphids are parthenogenetic and ovoviviparous during the long days of spring and summer but, during the short days of fall, males are stimulated to appear, fertilization takes place, and overwintering eggs are laid on the trees.

The circadian rhythm of some insect species (state of activity over a 24-hr period) has been correlated with susceptibility to insecticides, e.g., Triboleum is most effectively killed at 12 midnight.

EXAMPLES OF SOME INSECT LIFE HISTORIES

Mexican Bean Beetle (Epilachna varivestis)

Distribution.—Ranges over nearly all the United States.

Plants Attacked.—Beans of all types, cowpeas, soybeans, and beggar-ticks. May also feed on other legumes when infestations are heavy and food is scarce.

Injury.—Both larvae and adults feed mainly on the undersides of leaves (Figs. 3.3 and 3.4). The larvae usually leave the upper epidermis intact. Complete destruction of foliage, pods, and parts of stems follows heavy infestation.

Life History.—See Fig. 3.5. On the average it takes about a month for one complete generation from egg to adult. Adults usually arrive in fields as soon as the first planted beans have their first pair of leaves.

Grasshoppers

Species Concerned.—Rocky mountain grasshopper (*Melanoplus spretus*), the differential grasshopper (*M. differentialis*), the red-legged grasshopper (*M. femur-rubrum*), and the two-striped grasshopper (*M. bivittatus*).

Distribution.—Throughout North America.

Injury.—Consume foliage and other green plant parts.

Life History.—See Figs. 3.6 and 3.7. In some of the less important species the eggs hatch in the early fall and the young nymphs hibernate.

FIG. 3.3. MEXICAN BEAN BEETLE; ADULTS AND EGGS ON THE
UNDERSIDE OF A LEAF

FIG. 3.4. MEXICAN BEAN BEETLE; LARVAE, PUPAE, AND INJURY TO
LEAF

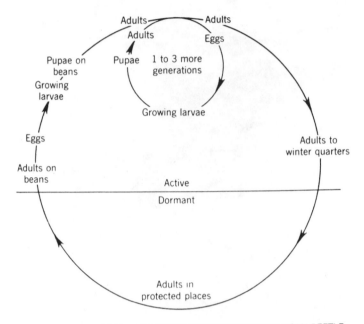

FIG. 3.5. THE SEASONAL LIFE HISTORY OF THE MEXICAN BEAN BEETLE

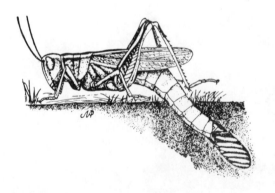

(From Col. Agr. Expt. Sta. Circ. 89-A)

FIG. 3.6. GRASSHOPPER LAYING EGGS IN THE GROUND

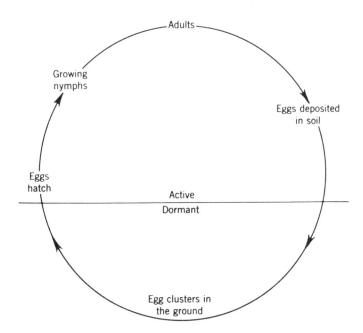

FIG. 3.7. THE LIFE HISTORY OF THE DIFFERENTIAL GRASSHOPPER

Rosy Apple Aphid (Dysaphis plantaginea)

Distribution.—General throughout the United States and Canada in the apple-growing regions.

Plants Attacked.—Apple, pear, thorn, and *Sorbus.* Narrow-leaved plantain is alternate host for some generations during the summer.

Injury.—Cause leaves to curl, stunt growth; apples remain small and malformed. Aphids excrete honeydew on which a sooty mold fungus grows.

Life History.—See Figs. 2.1A, 3.8, 3.9 and 3.10. About one generation is completed every 2 weeks in warm weather.

San Jose Scale (Quadraspidiotus perniciosus)

Distribution.—Throughout the United States and Canada where fruits are grown.

Plants Attacked.—Most fruit and some shade and forest trees.

Injury.—Sucking either weakens or kills plants or plant parts, or it disfigures and discolors fruits. Infested trees decrease in vigor, and foliage becomes thin and yellowed from presence of scales upon it. Fruit often with small, red, inflamed areas surrounding each scale.

FIG. 3.8. NEWLY HATCHED APHIDS ON AN APPLE BUD IN THE GREEN-TIP
STAGE OF DEVELOPMENT

A   B   C

(Redrawn with modifications from Mills and Dewey, Cornell Ext. Bull. 711)

FIG. 3.9. THE CORNICLES OF THREE SPECIES OF NEWLY HATCHED APHIDS ON
APPLES

A, Apple aphid; B, Rosy apple aphid; C, Apple grain aphid.

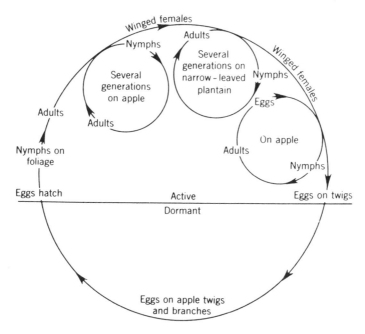

FIG. 3.10. THE LIFE HISTORY OF THE ROSY APPLE APHID

Life History.—See Figs. 3.11 and 3.12. Many scale species overwinter in the egg stage. Scales in tropical regions breed continuously. In San Jose scale there is an overlapping of generations, because females produce living young for about a month.

Onion Thrips (Thrips tabaci)

Distribution.—Throughout the United States and Canada.

Plants Attacked.—Onions, nearly all other garden crops, also many weeds, ornamental flowers, and some field crops.

Injury.—By rasping foliage and sucking sap, they produce grayish mottling of foliage which later may become distorted, turn rusty, and drop off or fall over. Plants may be killed or severely stunted (Fig. 3.13).

Life History.—See Figs. 3.14 and 3.15. Generations overlap each other so that all stages are generally found on plants at the same time. The metamorphosis is intermediate between gradual and complete with an inactive fourth instar similar to a pupal stage.

Peachtree Borer (Sanninoidea exitiosa)

Distribution.—Throughout the United States and British Columbia.

Plants Attacked.—Peach, wild and cultivated cherry, plum, prune,

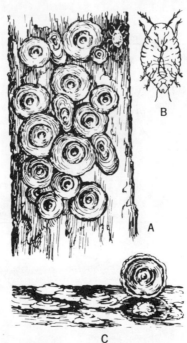

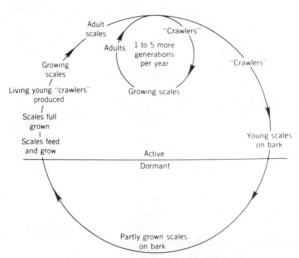

(Redrawn from Quaintance, U.S. Dept. Agr. Farmers' Bull. 650)

FIG. 3.11. SAN JOSE SCALE

A, Scales on limb; B, "Crawler" enlarged; C, Scale lifted to show insect underneath.

FIG. 3.12. THE LIFE HISTORY OF THE SAN JOSE SCALE (QUADRASPIDIOTUS PERNICIOSUS)

FIG. 3.13. ONION THRIPS INJURY ON ONION FOLIAGE

nectarine, apricot, almond, and some ornamentals of the genus *Prunus*.

Injury.—Larvae feed on inner bark and cambium above and below soil line. May completely girdle trees and kill them or weaken them and make them unproductive. Attack characterized by masses of gum mixed with frass from trunks near soil line (Fig. 3.16).

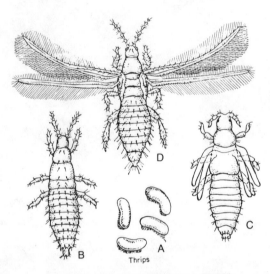

FIG. 3.14. GROWTH STAGES OF A THRIPS
A, Eggs; B, Nymph; C, Pupa; D, Adult.

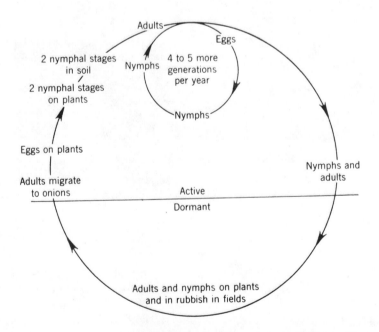

FIG. 3.15. THE LIFE HISTORY OF THE ONION THRIPS *(THRIPS TABACI)*

FIG. 3.16. PEACH TREE BORER INJURY

Life History.—See Fig. 3.17. The California peach tree borer (*Conopia opalescens*) has a similar life history. Some larvae do not pupate until the second summer.

European Corn Borer (Ostrinia nubilalis)

Distribution.—Over most of the eastern corn-growing areas of North America as far west as Iowa.

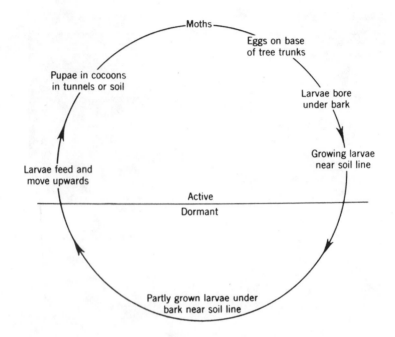

FIG. 3.17. THE LIFE HISTORY OF THE PEACH TREE BORER
(SANNINOIDEA EXITIOSA)

Plants Attacked.—Most herbaceous stemmed plants large enough for the larvae to tunnel into. Preferred hosts are corn, dahlias, gladioli, and weeds such as dock, cocklebur, pigweed, and smartweed. More than 200 host plants known.

Injury.—Larvae bore into stems and branches, weakening them and causing them to wilt or break over. On corn they attack the ear shanks, causing the ears to drop off, and also attack ears, reducing the crop. Sweet corn is made unsalable.

Life History.—See Figs. 3.18 and 2.1F. One, two, or more generations a year depending on the region. The southwestern corn borer (*Diatraea grandiosella*) has a similar life history.

Codling Moth (Laspeyresia pomonella)

Distribution.—Throughout apple-growing regions of North America.

Plants Attacked.—Fruit of apple, pear, quince, English walnut, wild haw, and crab apples.

Injury.—Larvae are the cause of "wormy" fruit or seeds and of premature dropping of fruit.

Life History.—See Figs. 3.19 and 3.20.

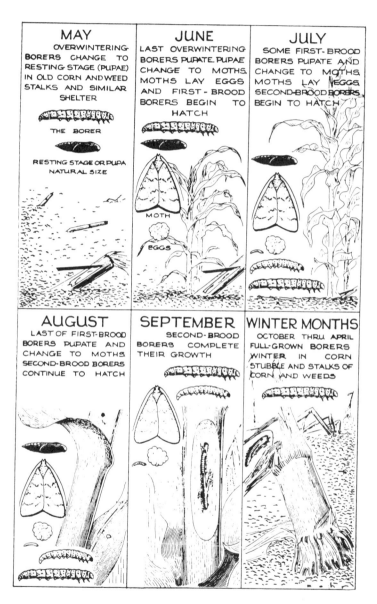

(From Univ. of Ill. Circ. 637)

FIG. 3.18. THE LIFE HISTORY OF THE EUROPEAN CORN
BORER (OSTRINIA NUBILALIS)

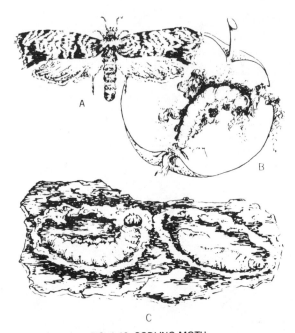

FIG. 3.19. CODLING MOTH

A, Adult; B, Larva in apple; C, Larva and pupa in hibernating cells under bark.

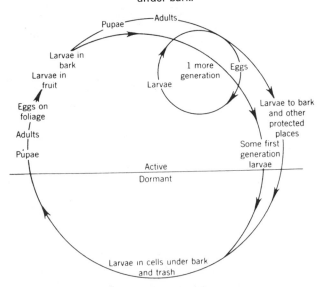

FIG. 3.20. THE LIFE HISTORY OF THE CODLING MOTH
(LASPEYRESIA POMONELLA)

Japanese Beetle (Popillia japonica)

Distribution.—Throughout eastern United States and in isolated colonies west to Missouri.

Plants Attacked.—Larvae feed on roots of grasses and of many other plants. Adults feed on foliage and fruit of most deciduous plants including many weeds.

Injury.—Larvae feeding on grass and vegetable roots destroy the plants (Figs. 3.21 and 3.22). Produce dead areas in lawns. Adults defoliate susceptible trees, shrubs, and herbaceous plants, also destroy ripening fruit. Larval stage of other white grubs may extend up to 3 years.

Life History.—See Fig. 3.23.

FIG. 3.21. VARIOUS SPECIES OF WHITE GRUBS IN THE SOIL

| Oriental Beetle | Japanese Beetle | Asiatic Garden Beetle | White Grub |

FIG. 3.22. ABDOMINAL SEGMENTS SHOWING DISTINGUISHING HAIR PATTERNS OF VARIOUS "WHITE GRUBS"

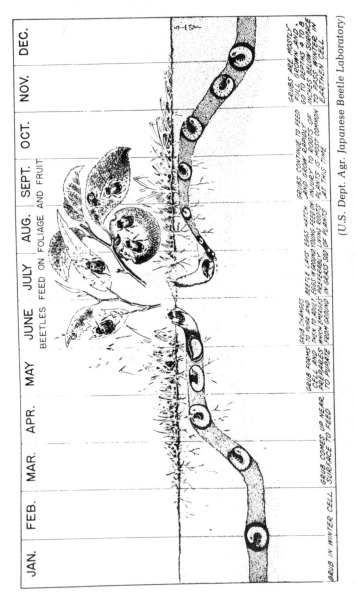

(U.S. Dept. Agr. Japanese Beetle Laboratory)

FIG. 3.23. THE LIFE HISTORY OF THE JAPANESE BEETLE

Cabbage Maggot (Hylemya brassicae)

Distribution.—Northern United States and Canada.

Plants Attacked.—All cruciferous plants such as cabbage, cauliflower, radish, turnip, and weeds belonging to the mustard group.

Injury.—Maggots feed on small fibrous roots and bore into larger ones. Plants stop growth, appear sickly, wilt in sun, and usually die or produce no crop.

Life History.—See Figs. 3.24 and 3.25. Onion maggot, cabbage maggot and seed corn maggot have similar life histories and appearance.

Wireworms (Family Elateridae)

Distribution.—Throughout North America.

Plants Attacked.—Roots of most plants in the grass family, such as corn and wheat; also clovers, beans, potatoes, and beets.

Injury.—Attack planted seed and prevent germination; feed on roots and tubers, often boring into them or into the underground parts of plant stalks, killing young plants.

Life History.—See Figs. 3.26 and 3.27. Life history varies in different species from 2–6 years.

FIG. 3.24. CABBAGE MAGGOT FLY

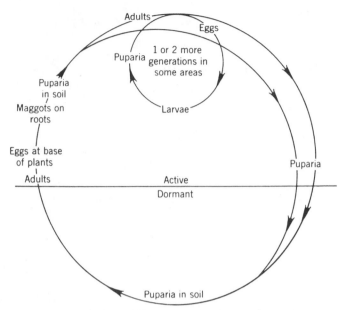

FIG. 3.25. THE LIFE HISTORY OF THE CABBAGE MAGGOT *(HYLEMYA BRASSICAE)*

FIG. 3.26. WIREWORM AND ADULT CLICK BEETLE

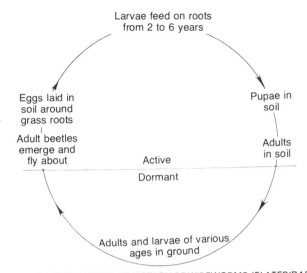

FIG. 3.27. THE GENERALIZED LIFE HISTORY OF WIREWORMS *(ELATERIDAE)*

TABLE 3.1

SOME IMPORTANT PESTS THAT CHEW OFF EXTERNAL PLANT PARTS

Insect	Scientific Name	Host Plants
Alfalfa weevil	*Hypera postica*	Alfalfa, clover
Blister beetles	Meloidae family	General feeders
Cabbage worms	*Pieris spp.*	Crucifers
Cankerworms	Geometridae family	Deciduous trees and shrubs
Colorado potato beetle	*Leptinotarsa decemlineata*	Potatoes, egg-plants, tomatoes
Cotton leafworm	*Alabama argillacea*	Cotton
Crickets	Gryllidae family	Grasses, grains, other plants and plant products
Cucumber beetles	*Diabrotica spp.*	Cucurbit family and some flowers
Cutworms	Noctuidae family	Nearly all herbaceous plants
Eastern tent cater-pillar	*Malacosoma americanum*	Apple, peach, plum, cherry, prune
Fall armyworm	*Spodoptera frugiperda*	Grasses, grains, vegetables, legumes, cotton, tobacco
Flea bettles	Chrysomelidae family	Vegetables, flowers, some fruits
Gypsy moth	*Porthetria dispar*	General feeder on deciduous trees
Japanese beetle	*Popillia japonica*	General feeder
Leaf rollers	Tortricidae family	Deciduous fruits and many ornamentals
Mexican bean beetle	*Epilachna varivestis*	Beans and other legumes
Webworms	Arctiidae family	Many deciduous and evergreen plants

TABLE 3.2

SOME IMPORTANT PESTS THAT SUCK ON PLANT PARTS

Insect	Scientific Name	Host Plants
Aphids	Aphididae family	Nearly all plant species
Chinch bug	Blissus spp.	Grass family, corn, grains
Harlequin bug	Murgantia histrionica	Cruciferous plants, many garden crops, fruits, weeds
Lace bugs	Tingidae family	Foliage of many trees and shrubs
Leafhoppers	Cicadellidae family	Numerous trees, shrubs, vegetables, flowers
Mealybugs	Pseudococcus spp.	All citrus, many ornamental and greenhouse plants
Psyllids	Psyllidae (Chermidae) family	Pear, boxwood, broad-leaved evergreens, many trees
Scale insects	Coccidae family	Numerous trees, shrubs, perennial greenhouse ornamentals
Spittle bugs	Cercopidae family	Grasses, legumes, trees, some fruits
Squash bug	Anasa tristis	Cucurbits
Tarnished plant bug	Lygus spp.	Many species of trees and herbaceous plants; alfalfa seed production reduced
Thrips	Thysanoptera order	Citrus, many deciduous trees, fruits, ornamentals, greenhouse crops
Treehoppers	Membracidae family	Numerous woody trees, shrubs
Whiteflies	Aleyrodidae family	Various tropical and greenhouse plants, laurel, rhododendron, azalea, etc.

TABLE 3.3

SOME IMPORTANT BORERS

Insect	Scientific Name	Host Plants
	Fruit, Nut, Seed, and Bud Borers	
Apple maggot	Rhagoletis pomonella	Apples, blueberries, wild crabs, haws
Boll weevil	Anthonomus grandis	Cotton
Clover seed midge	Dasyneura leguminicola	Red clover
Corn earworm or tomato fruit-worm	Heliothis zea	Corn, tomato, tobacco, cotton, etc.
Oriental fruit moth	Grapholitha molesta	Peach, quince, apple, pear, apricot, plum, etc.
Pickle and melon worms	Diaphania spp.	Most cucurbits
Pink bollworm	Pectinophora gossypiella	Cotton
Plum curculio	Conotrachelus nenuphar	Most pome and stone fruits
Spruce bud-worm	Choristoneura fumiferana	Spruce and balsam

TABLE 3.3 (*Continued*)

Insect	Scientific Name	Host Plants
Wood Borers		
Bark beetles	Scolytidae family	Many evergreen and deciduous trees
Carpenter-worm	*Prionoxystus robiniae*	Oaks, maple, cotton-wood, willow, chestnut
Flat headed borers	Buprestidae family	Many evergreen and deciduous trees
Leopard moth	*Zeuzera pyrina*	Elms, maples, etc.
Round headed borers	Cerambycidae family	Some shade and forest trees
White-pine weevil	*Pissodes strobi*	Pines and spruce
Herbaceous Plant Borers		
Corn billbugs	*Calendra spp.*	Corn, rice, large swamp grasses
Hessian fly	*Mayetiola destructor*	Wheat
Squash vine borer	*Melittia cucurbitae*	Squashes, at times other cucurbits
Stalk borer	*Papaipema nebris*	Most herbaceous plants
Sugarcane borer	*Diatraea saccharalis*	Sugarcane, corn, rice
Wheat stem maggot	*Meromyza americana*	Small grains, bluegrass, other grasses
Wheat stem sawfly	*Cephus cinctus*	Small grains, some native grasses
Leaf Miners		
Apple leaf trumpet miner	*Tischeria malifoliella*	Apples
Birch leaf miner	*Fenusa pusilla*	Birch
Boxwood leaf miner	*Monarthropalpus buxi*	Boxwood
Chrysanthemum leaf miner	*Phytomyza syngenesiae*	Chrysanthemums, Cineraria
Holly leaf miner	*Phytomyza ilicis*	Holly
Locust leaf miner	*Odontota dorsalis*	Black locust
Spinach leaf miner	*Pegomya hyoscyami*	Spinach, beets, Swiss chard

TABLE 3.4

SOME IMPORTANT SUBTERRANEAN PESTS

Insect	Scientific Name	Host Plants
Billbugs	Calendra spp.	Corn, grasses, grains, peanuts
Black vine weevil	Otiorhynchus sulcatus	Cylamen, begonia, yews, azalea
Corn root aphid	Anuraphis maidi-radicis	Corn, cotton, some grasses, weeds
Corn root-worms	Diabrotica spp.	Corn, grasses, many cultivated crops
Onion maggot	Hylemya antiqua	Onions
Seed-corn maggot	H. platura	Seed of corn, beans, lima beans, etc.
Strawberry rootworm	Chrysomelidae family	Strawberry, raspberry, grape, rose
Western grape rootworm	Bromius obscurus	Grapes and related plants
White-fringed beetles	Graphognathus spp.	Peanuts, corn, sugar cane, cotton, cabbage, sweet potatoes, etc.
White grubs	Scarabaeidae family	Grasses, many vegetables
Woolly apple aphid	Eriosoma lanigerum	Apple, pear, hawthorne, elm, mountain ash

TABLE 3.5

SOME IMPORTANT GALL-FORMING INSECTS

Insect	Scientific Name	Host Plants
Blister mites	Eriophyes spp.	Ornamentals, fruits, nuts
Chrysanthemum gall midge	Rhopalomyia chrysanthemi	Greenhouse chrysanthemums
Elm cockscomb gall	Colopha ulmicola	Elm
Grape phylloxera	Daktulosphaira vitifoliae	Grape
Horned oak gall	Callirhytis cornigera	Pin oak
Red-necked cane borer	Agrilis ruficollis	Raspberry, blackberry, dewberry
Spruce gall aphids	Adelges spp.	Norway spruce, blue spruces, Douglas fir

QUESTIONS FOR DISCUSSION

1. Of what importance is it to a grower to know the life history of his crop pests?
2. What factors would lead to the presence of all stages of an insect on a plant at the same time?
3. Which pest would be easier to control: one that has definite broods with little overlapping of stages or one in which there is so much overlapping of stages that no definite broods can be recognized?
4. Present reasons why most northern insects don't freeze to death in the winter even though temperatures drop far below freezing.
5. What factors operate to produce hibernation? Estivation?
6. Would you expect an insect to have the same number of generations in northern states as in southern states?
7. Since insects are cold-blooded, why aren't temperatures that are optimum for one optimum for all?
8. What are the differences between the incubation of chicken eggs and of insect eggs?
9. Tropical or greenhouse insects fail to survive freezing temperatures. Why?
10. Is dormancy in plants similar to hibernation in insects? Give reasons.
11. What conditions produce high insect mortality in the summer? In the winter?
12. Where do insects that live in dry food products get their moisture?
13. What insects tend to flourish during wet seasons? Dry seasons?
14. How does light regulate insect activities?
15. Study each of the life histories presented in this chapter and pick out the possible points where each pest may be attacked.
16. Why is accurate timing of control measures dependent on a knowledge of the life history of a pest?
17. Discuss the importance of correct timing of control practices with foliage-chewing insects. With sucking insects. With boring insects. With subterranean insects.
18. Is it good policy for a grower to have to depend on some agricultural agency to inform him of a pest's appearance and of when control measures should be applied?
19. Why is the Mexican bean beetle a bad pest on bean plantings throughout the summer?
20. Grasshopper plagues are no longer common in the United States. Why?
21. Why is control usually directed against the larvae of moths and butterflies and not against the adults?

22. In what stages are scale insects easiest to control?
23. Why are aphids potentially such destructive pests?
24. Control applications for most sucking pests should be directed to the undersides of foliage. Why?
25. Why are control measures for boring insects usually directed against the adults or the newly hatched larvae?
26. Why is correct timing very important in the control of leaf miners?
27. In what stages are root-feeding insects subject to attack?
28. Plant galls are produced by the irritation of insects. What is the function of the gall?

Chapter 4

The Classification of Insects

THE NAMING OF ORGANISMS

The grouping of organisms, referred to as classification, has made scientific progress in biology possible. An organism must be recognized and identified correctly before one can deal effectively with it from the standpoint either of destroying it or of breeding it for useful purposes. Animal and plant classification is based on the *Binomial System of Nomenclature* in which each organism is known by two Latin names, the genus name and the species name, somewhat similar to our family name and given name. For example, the name of the squash bug is *Anasa tristis*, *Anasa* being the genus (to which a number of species may belong) and *tristis*, being the species to which the squash bug belongs. These two names constitute the scientific name of the insect, which is recognized throughout the world, as opposed to the common name which may differ in each language and locality.

Definition of a Species

A species may be defined as a "reproductively isolated, naturally distinct group of organisms"; or a group of organisms sufficiently alike to have had the same parents. These definitions are not foolproof, and exceptions to them may be found. Although the species is the working unit of the biologist, genetic modifications within a species frequently occur, especially in the plant kingdom, so that subspecies, variety, race, or strain names are employed in addition to the genus and species names. In scientific writing the scientific name also includes the name of the man that originally described the species.

Classification of Organisms

Starting with the species as the smallest unit of classification, a number of species that are closely related are grouped together to form a *genus*; genera (plural of genus) with similar characteristics are combined to form a *family*; large families are sometimes divided into sub-families, or families are combined to form superfamilies; similar families are combined to form an *order*; similar orders to form a *class*; and similar classes to form a *phylum* (animals) or a *division* (plants). The phyla are all combined into the *animal kingdom*; the plant divisions are combined into the *plant kingdom*. The two kingdoms include all the living organisms that are or were known to exist on this earth. Different, closely related insects may be classified alike up to the species name and may have the same generic name. As an example we have the native cabbage butterfly, *Pieris aleracea*, and the imported cabbage butterfly, *Pieris rapae*. Different species generally cannot interbreed.

TABLE 4.1

EXAMPLES OF INSECT CLASSIFICATION

	I	II
Kingdom	Animal	Animal
Phylum	Arthropoda	Arthropoda
Class	Hexapoda	Hexapoda
Order	Lepidoptera	Hemiptera
Family	Olethreutidae	Coreidae
Genus	*Laspeyresia*	*Anasa*
Species	*pomonella*	*tristis*
Scientific name	*Laspeyresia pomonella*	*Anasa tristis*
Common name	Codling moth	Squash bug

Common names of arthropods are generally useful in local or regional areas, but because these may differ in different regions, countries or languages scientific names are indispensable for accurate identification and use of references throughout the world. To help standardize common names, The Entomological Society of America publishes a list of approved common and scientific names of pest and beneficial arthropods.

The class Hexapoda in the phylum Arthropoda is divided into some 26 orders on the basis of such characteristics as the absence or presence of wings; wing texture, venation and number; the type of mouthparts; and the type of metamorphosis. Not all entomologists are in agreement on the limiting characteristics of each order, and some have decided to further divide some orders to create new ones. The author is retaining the established group of recognized orders.

All insects belong to the class Hexapoda in the phylum Arthropoda. The Hexapoda are divided into 22 more important and several lesser orders, each with many families and genera. The classification and characteristics of the agriculturally important groups of the animal kingdom are shown in Table 4.2.

TABLE 4.2

CLASSIFICATION OF THE AGRICULTURALLY IMPORTANT GROUPS
OF THE ANIMAL KINGDOM

Phylum	Class	Characteristics	Examples
		Vertebrates (Animals with a Backbone)	
Chordata	Mammalia*	Hairy, four-footed, milk-secreting	Mammals such as man, cow, mice, rabbits
	Aves*	Wings and feathers	Birds such as chickens, ducks, robins, hawks
	Reptilia	Cold-blooded, scaly, air-breathing	Reptiles such as snakes, lizards, turtles
	Amphibia	Cold-blooded, moist, soft-skinned; young aquatic with gills, adults with lungs	Amphibians such as frogs, toads, salamanders
	Pisces	Cold-blooded with scales and fins, aquatic, gill-breathing	Fish such as salmon, bass trout
		Invertebrates (Animals without a Backbone)	
Arthropoda		Chitinous exoskeleton, jointed appendages, segmented body	
	Hexapoda*	Head, thorax, abdomen, 3 pairs legs, wings	All true insects: bugs, beetles, moths, etc.
	Diplopoda*	Head and abdomen, many-segmented, 2 pairs legs per segment	Millipedes
	Chilopoda	Flattened body, head and abdomen, 1 pair legs per body segment, pair poison fangs on first body segment	Centipedes
	Arachnida*	Fused head and thorax (cephalothorax), 4 pairs walking legs, abdomen distinct, unsegmented, no antennae, only simple eyes; no evident body divisions in mites and ticks	Spiders, mites, ticks, scorpions
	Crustacea*	Hard, limey, chitinous exoskeleton, 2 pairs antennae, cephalothorax usually with biramous appendages	Crayfish, crabs, sowbugs, lobsters
	Symphyla*	Head and abdomen, 12 pairs legs, (tiny, enlongated arthropods)	Symphilids or symphilans
Mollusca*		Soft-bodied, non-segmented, no jointed appendages; body may be enclosed in calcareous shell	Snails, slugs, clams, oysters
Annelida*		Elongated, segmented worms, no appendages	Earthworms, leeches
Nemata*		Tiny, elongated, cylindrical, unsegmented with tough cuticle; many species microscopic	Nematode worms, hookworms, trichinella
Platyhelminthes		Flattened, usually unsegmented worms, many microscopic	Flatworms, flukes, tape-worms
Protozoa		Single-celled animals, often with cilia	Ameba, paramecia, trypanosomes

*Includes forms injurious to plants.

THE ORDERS AND FAMILIES OF INSECTS IMPORTANT IN AGRICULTURE

Insect Orders with No Metamorphosis

Collembola.—Springtails. Springtails are minute wingless insects; usually white, gray or yellow in color; with a forked spring (**furcula**), used in jumping, folded forward under the posterior end of the abdomen. Mouthparts are modified chewing and withdrawn into the head. Most feed on decaying plant material but a few species, such as the garden springtail, pepper the foliage of seedling plants with tiny holes and cause damage in mushroom houses.

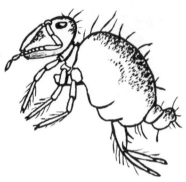

Thysanura.—Bristletails, Silverfish, Firebrats. These are active small- to medium-sized wingless chewing insects with elongate tapering bodies gray, tan or brown in color; with three long tail-like appendages extending from the posterior end of the abdomen; and with vestigial legs (**styli**) on some abdominal segments. The body is generally covered with scales. They are usually found in buildings near heat (firebrats) or in cool damp areas (silverfish). They are destructive to all types of starch-containing materials in homes and storage buildings.

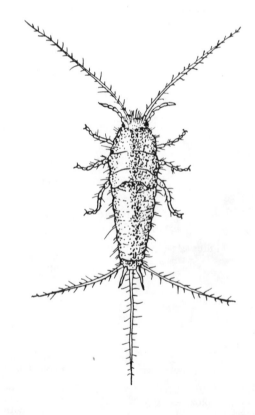

Insect Orders with a Gradual or Incomplete Metamorphosis

Ephemeroptera.—Mayflies. Mayflies are small- to medium-sized four-winged insects with vestigial mouthparts, short bristlelike antennae and with two or three long thread-like segmented tails. The forewings are large triangular and many-veined; the hind wings (lacking in some species) are small and rounded. The wings are held in a vertical position when at rest. The immature stages (naiads) are aquatic with leaflike or plumose abdominal gills. They feed on algae and other plant life. Mayflies are considered beneficial because they serve as an important source of food for many freshwater fish.

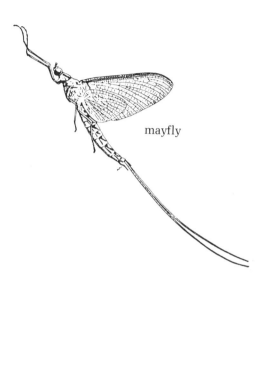

mayfly

Odonata.—Damselflies, Dragonflies. These are medium-sized to large chewing insects with four elongate, membraneous many-veined wings, large compound eyes, short bristlelike antennae and long slender abdomens. The nymphs (naiads) are aquatic and breathe by means of three leaflike anal gills (damselflies) (Fig. 2.3G) or by internal ridgelike rectal gills (dragonflies). The naiads possess a long, hinged and extendible labium modified to grasp small prey. The wings of damselflies are held together above the body when at rest; those of dragonflies are held extended separately at right angles to the

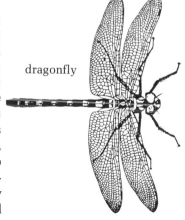

dragonfly

body. The adults are also predacious and catch their prey (midges, mosquitoes, etc.) on the wing with their legs. This order is considered beneficial.

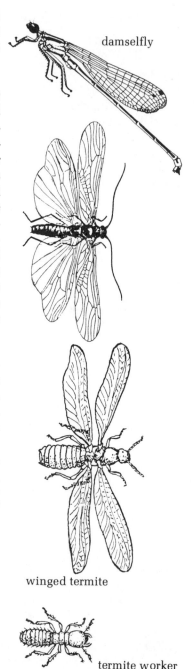

damselfly

Plecoptera.—Stoneflies. Stoneflies are mostly medium-sized somewhat flattened insects with long, slender many-segmented antennae and two pairs of multi-veined membraneous wings held flat over the abdomen when at rest. The front wings are slightly longer than the lobed hind wings. Cerci may vary in length and are multi-segmented. Mouthparts are chewing or vestigial. Naiads are flattened, elongate with cerci and usually with branched gills on the thorax or first two or three abdominal segments. The naiads may be herbivorous, predaceous or omnivorous and are considered important food for many fish.

Isoptera.—Termites. Termites are generally small pale-colored social insects that live in colonies with a highly developed caste system made up of a wingless queen, workers (nymphs and sterile adults) and soldiers; and a winged reproductive caste made up of darkly pigmented males (kings) and queens and supplementary reproductives usually with short wing pads. The two pairs of wings are much longer than the body and are similar in size, shape and venation. These wings break off during or after the swarming, dispersal and mating flights. Termites have chewing mouthparts, short, segmented string-like antennae, short segmented cerci and a thick waist (in contrast to the thin waist in ants). They feed on cellulose materials which they digest with the aid of symbiotic protozoa in

winged termite

termite worker

their intestines. They are serious pests of wooden structures throughout the world. Some are subterranean and others (drywood termites) can live above ground without soil contact.

Dermaptera.—Earwigs. Earwigs are medium-sized slender, brown to black chewing insects characterized by large unsegmented forcepslike cerci and pad-like short and leathery front wings that cover the larger membraneous hind wings which are folded underneath. Earwigs are nocturnal insects. One species, the European earwig, may be destructive to vegetable crops, fruits and ornamentals.

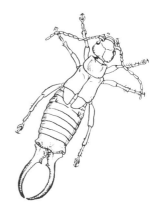

Psocoptera.—Psocids, Booklice, Barklice. These are small four-winged or wingless insects with chewing mouthparts and large or reduced compound eyes. The wings when present have few veins and are held rooflike over the abdomen when at rest. The antennae are moderately long and cerci are lacking. The indoor species are generally wingless; the outdoor species, found on or under bark, are mostly winged. Psocids feed on molds, fungi, algae, cereals, grains, paper, glue and starches. Under high humidity conditions they may become numerous indoors and become destructive.

Thysanoptera.—Thrips (Fig. 3.14). These are tiny to small slim insects with rasping-sucking mouthparts in a cone-like beak attached far back on the ventral side of the head. They are characterized by four long narrow wings fringed with long hairs and by a one- or two-segmented tarsi tipped with bladder-like swellings. Their metamorphosis approaches com-

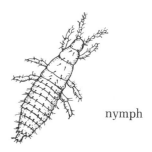

nymph

pleteness because there is a quiescent pupal-like instar preceding the adult stage. Some species cause serious injury to plants by attacking stems, leaves and blossoms; a few are predatory on mites and small insects.

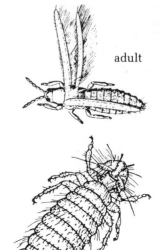

adult

Mallophaga.—Chewing lice. This order consists of small flattened wingless ectoparasites with chewing mouthparts; a large, wider than long, head; and short antennae (in some species concealed in grooves). They feed on bits of hair, feathers, dead skin or scabs causing irritation and annoyance and producing under heavy infestation an unhealthy appearance or even death. Domestic animals, especially poultry, may suffer from their attacks (see Table 4.3).

poultry body louse

Anoplura.—Sucking lice. These lice are minute to small, wingless, blood-sucking external parasites with head narrower than long, with piercing-sucking mouthparts retracted into the head, and with tiny eyes. Their short sturdy legs have a single-segmented tarsus and one large claw adapted for clinging to the hairs of the host. Several species are ectoparasites on domestic animals and two (the head and body louse and the crab louse) infest humans. These lice are very irritating to their hosts and in addition are capable of transmitting such human diseases as relapsing fever, trench fever and epidemic typhus through their feces or crushed bodies.

hog louse

Orthoptera.—Grasshoppers, Crickets, Mantids, Walking sticks, Katy-dids and Cockroaches (Fig. 4.1). The group consists of medium- to large-sized robust to slim, elongate insects with chewing mouthparts

TABLE 4.3

THE ECONOMICALLY IMPORTANT FAMILIES OF MALLOPHAGA

Family	Distinguishing Characteristics	Importance
Menoponidae Poultry body lice (see p. 66)	Antennae more or less clubbed, in grooves on sides of head. Head broadly triangular, expanded behind eyes	Serious ectoparasites on poultry and birds
Philopteridae Feather-chewing lice	Antennae filiform, exposed, 5-segmented. Tarsi with 2 claws	Serious ectoparasites on poultry and birds
Trichodectidae Mammal-chewing lice	Antennae 3-segmented exposed. Tarsi with one claw	Serious ectoparasites on mammals

and usually with two pairs (sometimes reduced or absent) many-veined wings: the forewings are elongate parchment-like **(tegmina)**; the hindwings are membraneous, larger and folded fan-like when at rest. The compound eyes are large, the legs are usually long and cerci are present. Short to long ovipositors and singing are characteristic of many members. This group is mainly phytophagus but some are omnivorous and some are useful predators.

Hemiptera.—True Bugs (Fig. 4.2). The true bugs are small to large oval to elongate insects with piercing-sucking mouthparts in a segmented beak attached to the anterior end of the head and extending backward between the coxae when not in use. Adults have two pairs of wings or are wingless. When winged their bases are separated by a large triangular scutellum. The forewings are anteriorly thickened, distally membraneous hemelytra that overlap on the abdomen. This group includes many destructive plant pests; some aquatic insects; some animal and human ectoparasites that are capable of transmitting diseases; and some predators on other insects.

Homoptera.—Aphids, Cicadas, Leafhoppers, Scale Insects, Mealybugs, Whiteflies, Psyllids, Spittlebugs (Fig. 4.3). This diverse group, ranging from minute to large insects, has in common piercing-sucking mouthparts in a short beak originating from the back ventral side of the head and extending between the front coxae. Antennae are short bristlelike or threadlike with few segments; wings are present or absent; when present there are two membraneous pair usually held

TABLE 4.4

THE ECONOMICALLY IMPORTANT FAMILIES OF ORTHOPTERA

Family	Distinguishing Characteristics	Importance
Acrididae Shorthorn grasshoppers Fig. 4.1F	Medium- to large-sized, with antennae no longer than pronotum; with the tympanum on sides of first abdominal segment near base of hindwings and with short ovipositor	Phytophagus—capable of great destruction to crops
Blattidae **Cockroaches** Fig. 4.1C	Small to large insects with wings held flat, as long as or shorter than the abdomen. Body broad and flat with cerci. Head concealed under large pronotum, antennae long filiform	Omnivorous—feed on and contaminate with feces all types of stored food products
Gryllidae Crickets Tree crickets Mole crickets Fig. 4.1D, E	Oval robust (slender in tree crickets) chirping insects with wings held flat. Wings short, long or absent. Ovipositor needlelike or spearlike, cerci prominent. Front legs of mole crickets adapted for burrowing. Nocturnal	Phytophagous or omnivorous. Mole crickets may destroy seeds and seedlings
Mantidae Mantids Fig. 4.1A	Large elongate twig-like or leaflike bodies with long raptorial front legs, greatly lengthened thorax and moveable head	Beneficial—predaceous on other insects
Phasmidae Walking sticks Fig. 4.1B	Wingless, long stick-like bodies with long hairlike antennae. Mostly tropical species	Phytophagus—may do serious damage to trees when abundant
Tettigoniidae Longhorned grasshoppers Katydids Fig. 4.1G	Medium- to large-sized with long hairlike antennae and sword- or sicklelike ovipositor	Phytophagus—may harm foliage of ornamentals and crops

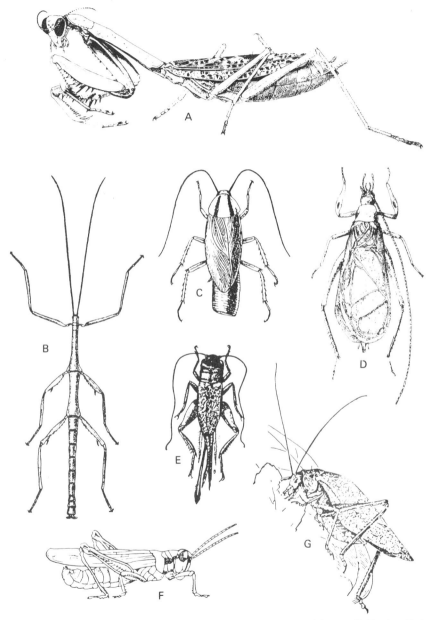

(A, G, courtesy University of California; B, E, after Essig;
C, F, courtesy U.S.D.A.; D, courtesy University of California)

FIG. 4.1. FAMILIES OF ORTHOPTERA

A, **Mantidae,** praying mantis; B, **Phasmidae,** walking stick; C, **Blattidae,** German roach
with egg case; D, **Gryllidae,** snowy tree cricket; E, **Gryllidae,** field cricket; F, **Acrididae,**
migratory grasshopper; G, **Tettigoniidae,** broad-winged katy-did.

sloping over the body. Some are sessile and scale-covered and some deposit a sweet excretion (**honeydew**) on foliage that is attractive to ants and on which a black sooty-mold fungus develops. This order is highly destructive to many crops and contains many vectors of destructive plant viral diseases.

TABLE 4.5

THE ECONOMICALLY IMPORTANT FAMILIES OF HEMIPTERA

Family	Distinguishing Characteristics	Importance
Cimicidae Bedbugs Fig. 4.2A	Small, broadly-oval flattened bodies with vestigial wings, 3-segmented beak, short filiform antennae. Nocturnal	Feed on blood of humans, animals, birds. Bites very irritating
Coreidae Coreid bugs Leaf-footed bugs Squash bugs Fig. 4.2B	Medium-sized usually elongate-oval, odiferous bugs with membraneous part of hemelytra many-veined. Hind tibiae flanged in leaf-footed bugs	Occasionally harmful to cucurbits, fruits, and trees. Few are predaceous
Lygaeidae Chinch bugs Seed bugs Fig. 4.2C	Small elongate or elongate-oval, sometimes colorful, with membraneous part of hemelytra with few simple open veins. Ocelli usually present. Long-winged and short-winged forms in some species	Feed on seeds and on plant juices of cereals and grasses. Injury may be serious
Miridae Plant bugs Leaf bugs Lygus bugs Fig. 4.2E	Small oval or elongate-oval, often strikingly marked bugs with a more or less triangular section (**cuneus**) in apical tip of opaque part of hemelytra, membraneous part with two closed cells	Feed on juices of wild and cultivated plants and fruits. Some species serious pests
Pentatomidae Stink bugs Shield bugs Harlequin bugs Fig. 4.2D	Medium-sized to large bugs with broadly oval shield-like shape; with small heads, 5-segmented antennae and large scutellum	Pests of vegetable crops, especially cabbage family. Some predaceous on insects

TABLE 4.5 *(Continued)*

Family	Distinguishing Characteristics	Importance
Pyrrhocoridae Red bugs Cotton stainers Fig. 4.2G	Medium-sized to large bugs, rounded to elongate with small heads and large eyes. Membrane of forewing with many-branched veins and crossveins. Some short-winged	Feed on seeds. Highly destructive to cotton and other crops
Tingidae Lace bugs Fig. 4.2F	Small flattened bugs with sculptured lacelike wings and broadly-flanged pronotum. Nymphs spiny	Feed on foliage of trees and shrubs. Cause discoloration and defoliation

Insect Orders with a Complete Metamorphosis

Coleoptera.—Beetles, Snout Beetles, Weevils (Figs. 4.4, 4.5). These are small to large hard-bodied insects varying in shape from rounded to elongated and in type of antennae. Adults have large compound eyes, chewing mouthparts and usually two pairs of wings which may be greatly reduced in some species. The forewings, elytra, may be thickened and leathery or hard and brittle. The hind pair are membraneous and fold up under the elytra when at rest. In some beetles the head is prolonged into a short or long snout with the mouthparts at the distal end. The larvae are also chewing and vary greatly in form, habitat and food requirements in different families. Some beetles are aquatic in both the larval and adult stages. Many beetles and their larvae are phytophagous; some are predaceous on other insects; and some feed on stored plant and animal products, fabrics or decaying organic matter. They are responsible for serious crop losses but some species are valuable as predators or scavengers.

Lepidoptera.—Moths, Butterflies (Figs. 4.6, 4.7). This group consists of small to large insects usually with two pairs of large colorful wings covered with overlapping pigmented scales. Their bodies are robust or elongated and usually covered with pigmented hairs or scales. Mouthparts consist of a long slim tube (vestigial in a few species) for sucking up (siphoning) liquids, mainly nectars of flowers. This tube is coiled up under the head like a spring when at rest. The antennae are knobbed in butterflies but of various forms, other than knobbed, in moths; the compound eyes are large and the legs relatively long.

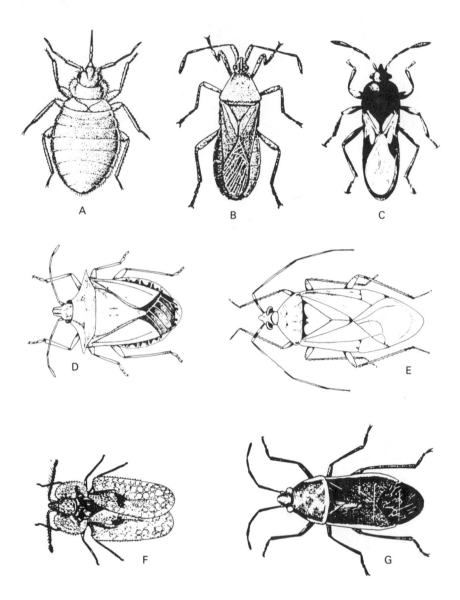

*(A, after Hermes; B, C, F, courtesy U.S.D.A.; D, E, courtesy
University of Arizona, College of Agriculture; G, after Essig)*

FIG. 4.2. FAMILIES OF HEMIPTERA

A, **Cimicidae,** bed bug; B, **Coreidae,** squash bug; C, **Lygaeidae,** chinch bug; D,
Pentatomidae, western brown cotton bug; E, **Miridae,** lygus bug; F, **Tingidae,** lace
bug; G, **Pyrrhocoridae,** borded plant bug.

TABLE 4.6

THE ECONOMICALLY IMPORTANT FAMILIES OF HOMOPTERA

Family	Distinguishing Characteristics	Importance
Aleyrodidae Whiteflies Fig. 4.3A	Minute white-winged, moth-like insects with wings covered with white powdery wax. Immature stages beyond the first instar yellow or black, spiney, sessile and scale-like. Produce honeydew	Can cause serious damage to citrus, azaleas, and greenhouse plants by sucking sap from foliage and producing honeydew
Aphididae Aphids Plantlice Fig. 4.3B	Small plump, soft-bodied winged or wingless insects with fairly long antennae and a pair of cornicles at posterior end of abdomen. Some species covered with wooly fibers. Produce honeydew	Destructive foliage pests of many different crops and plants. Some attack roots or produce galls
Cercopidae Froghoppers Spittle-bugs Fig. 4.3C	Small oval, robust, hopping insects with hind tibiae sparsely spined. Nymphs concealed in frothy masses of spittle	Attack grasses, herbaceous plants, and some pines. May cause serious stunting of plants
Cicadellidae Leafhoppers Fig. 4.3E	Small elongate wedge-shaped insects with wings held rooflike and hind tibiae with one or more rows of small spines. Nymphs crawl sidewise like crabs, may produce honeydew	Destructive pests of many different crops and plants—discolor foliage, stunt growth, and transmit many serious plant viral diseases
Cicadidae Periodical cicadas Dog-day cicadas Fig. 4.3J	Large robust insects with prominent eyes, 3 ocelli, short bristlelike antennae, and large membraneous wings held roof-like. Males sing, females with strong spear-like ovipositors. Nymphs suck sap from tree roots	Egg-laying punctures in twigs of trees and shrubs very injurious when broods appear

TABLE 4.6 *(Continued)*

Family	Distinguishing Characteristics	Importance
Coccidae Soft scales Tortoise scales Wax scales Fig. 4.3H	Females small, flattened, hemispherical or elongate-oval with cleft in posterior end. Bodies coated with hard or soft waxy material. Wingless, usually legless and sessile. Males with a single pair of wings. First instar nymphs active "crawlers"	Destructive to many trees shrubs and greenhouse plants. Weaken or kill plants, produce honeydew
Diaspididae Armored scales Hard scales Fig. 4.3D	Tiny small soft-bodied insects concealed under hard waxy removable shells. Flattened disc-like or oval bodies lacking appendages and distinct segmentation. First instar young active "crawlers"	Very serious pests of trees and shrubs
Membracidae Treehoppers Fig. 4.3F	Small insects with grotesque-shaped pronotum extending over abdomen, hump-backed appearance	Mainly damaging to twigs: egg-laying punctures stunt and distort growth
Pseudococcidae Mealybugs Fig. 4.3G	Small sluggish, wingless, oval insects, usually covered or fringed with white waxy or mealy secretion. Eggs laid in loose cottony masses	Serious damage to citrus, taxus and various greenhouse or tropical plants
Psyllidae Psyllids Jumping plantlice Fig. 4.3I	Small, resemble miniature cicadas, with strong jumping legs; hind femora enlarged and antennae relatively long. Nymphs may produce white waxy secretion or honeydew	Feed on plant juices, transmit yellows virus to vegetables of nightshade family. Few species produce galls

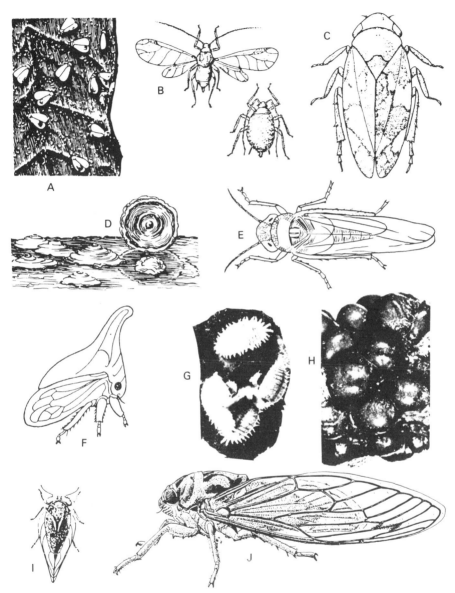

(A, B, D, E, F, courtesy U.S.D.A.; C, after Borer and DeLong; G, H, courtesy University of California, Riverside; I, courtesy Connecticut Agricultural Experiment Station; J, courtesy C.O. Mohr)

FIG. 4.3. FAMILIES OF HOMOPTERA

A, **Aleyrodidae,** greenhouse white fly; B, **Aphididae,** winged and wingless aphid; C, **Cercopidae,** froghopper; D, **Diaspididae,** San Jose scale; E, **Cicadellidae,** potato leafhopper; F, **Membracidae,** treehopper; G, **Pseudococcidae,** citrus mealybug; H, **Coccidae,** hemispherical scale; I, **Psyllidae,** pear psylla; J, **Cicadidae,** dog-day cicada.

TABLE 4.7

THE ECONOMICALLY IMPORTANT FAMILIES OF COLEOPTERA

Family	Distinguishing Characteristics	Importance

Buprestidae
Larvae: Flat-
   headed wood
   borers
Fig. 4.4A1, A2
	Usually medium to large-sized metallic colored beetles with hard flattened bodies, posteriorly tapered elytra, large eyes, and short serrate antennae. Thorax of larvae usually expanded and flattened	Larvae are wood-borers. Damage serious to trees and shrubs

Carabidae Predaceous ground beetles Fig. 4.4D	Minute to large hard-bodied elongate beetles with strong, pointed mandibles; filiform antennae, large eyes and dull to metallic colored bodies. Nocturnal	Both adults and larvae predaceous on insects. A few are phytophagus

Cerambycidae
Long-horned
   beetles
Larvae: Round-
   headed wood-
   borers
Fig. 4.4C1, C2
	Small to large usually colorful, elongate, cylindrical bodies usually with very long filiform or serrate antennae. Eyes notched near base of antennae	Larvae destructive wood-borers

Chrysomelidae Leaf beetles Tortoise beetles Flea beetles Larvae: Root-worms Fig. 4.4B1, B2	Generally small, oval, convex-bodied beetles with short filiform to clavate antennae. Usually brightly colored or metallic bodies	Adults and larvae phytophagus. Destructive to foliage of plants or roots of crops
Coccinellidae Lady beetles Fig. 4.4E1, E2	Small oval, almost hemispherical insects, usually with brightly colored and spotted elytra, short clavate antennae and 3 tarsal segments (Chrysomelidae appear to have 4)	Both adults and larvae valuable predators on soft-bodied insects (aphids, scales). Two species destructive, phytophagus (Mexican bean beetle, squash beetle)
Curculionidae Weevils, snout beetles Curculios Fig. 4.4F1, F2	Small to large beetles with head prolonged into a short or long snout with chewing mouthparts at distal end. Antennae geniculate and clubbed, arising from the snout. Larvae grublike, legless, with distinct head	Destructive: Adults drill holes in feeding or egg-laying in fruits, nuts, stems, seedpods and seeds. Larvae feed within plant tissues

TABLE 4.7 *(Continued)*

Family	Distinguishing Characteristics	Importance
Dermestidae Dermestid beetles Skin beetles Carpet beetles Fig. 4.5A1, A2	Small oval or hemispherical beetles with head concealed from above, clubbed antennae; elytra color-patterned by hairs or scales. Larvae brownish with long tail hairs or tufts	Adults feed on pollen or animal products. Larvae destructive to animal products: leather, furs, skins, museum specimens, woolens, silks, rugs and stored foods
Elateridae Click beetles Larvae: Wireworms Fig. 4.5B1, B2	Minute to medium-sized, with elongate, somewhat flattened bodies. Dark-colored or metallic (in tropics). Antennae filiform or serrate. Prothorax large with pointed posterior angles. Larvae: long, slim, cylindrical and hard-bodied	Larvae destructive to the roots of vegetables, cereals, and forage crops
Lyctidae Powder-post beetles Fig. 4.5F	Minute to small, with slender elongate brown to black bodies. Antennae short with two serrate segments at tip. Larvae white and grublike	Adults and larvae destructive to dry seasoned unpainted wood. Pepper wood with tiny holes, may reduce wood to fine powder
Meloidae Blister beetles Fig. 4.5C	Medium-sized. Body usually narrow and elongate with soft, flexible sometimes shortened elytra. Pronotum narrower than head and elytra	Some adult species important pests of vegetables. Larvae predatory on grasshopper eggs
Scarabaeidae June beetles May beetles Chafers Dung beetles Larvae: white grubs Fig. 4.5G1, G2	Small to large, robust elongate-oval, dull to brilliantly colored. Antennae short lamellate. Nocturnal. Larvae white grublike, C-shaped	Adults destructive to foliage and larvae serious root pests of many crops
Seblytidae Bark beetles Engraver beetles Shot-hole borers Fig. 4.5E1, E2	Minute to small compact cyclindrical and blunt-ended. Brownish to black. Larvae: tiny, white, grublike	Both adults and larvae destructive—mine under the bark or into the wood of forest or fruit trees
Tenebrionidae Darkling beetles Ground beetles Flour beetles Fig. 4.5D	Tiny to medium-sized beetles with hard, flattened or cyclindrical bodies. Brown to black in color. Antennae short filiform or clavate. Eyes usually notched. Larvae: Some similar to wireworms	Both adults and larvae destructive to stored cereals, grains, flour and dried fruits

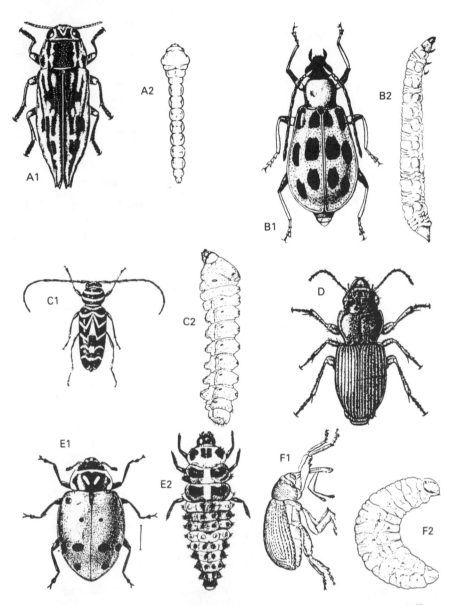

(Figs. A, B, C, E and F courtesy U.S.D.A.; Fig. B, courtesy Kansas State College)

FIG. 4.4. FAMILIES OF COLEOPTERA

A1, A2, **Buprestidae**, metallic wood borer and larva; B1, B2, **Chrysomelidae**, spotted cucumber beetle or southern corn rootworm; C1, C2, **Cerambycidae**, locust borer and larva; D, **Carabidae**, ground beetle; E1, E2, **Coccinellidae**, convergent lady beetle and larva; F1, F2, **Curculionidae**, cotton boll weevil and larva.

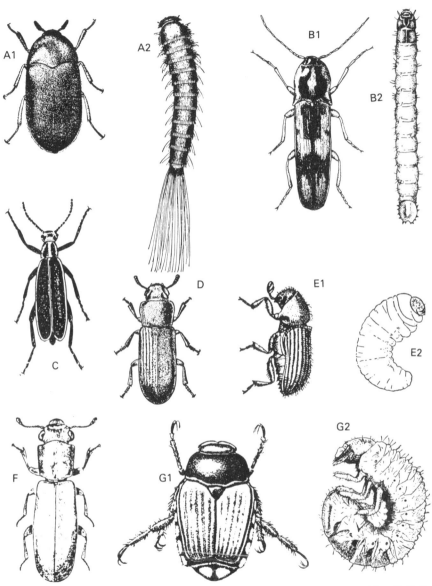

(Fig. A, courtesy Connecticut Agricultural Experiment Station; Figs. B, D, E, courtesy U.S.D.A.; Fig. F, after Arnett; Fig. C, courtesy U.S. Public Health)

FIG. 4.5. FAMILIES OF COLEOPTERA

A1, A2, **Dermestidae,** black carpet beetle and larva; B1, B2, **Elateridae,** southern corn wireworm; C, **Meloidae,** margined blister beetle; D, **Tenebrionidae,** confused flour beetle; E1, E2, **Scolytidae,** peach bark beetle and larva; F, **Lyctidae,** powder-post beetle; G1, G2, **Scarabaeidae,** Japanese beetle and larva.

TABLE 4.8

THE ECONOMICALLY IMPORTANT FAMILIES OF LEPIDOPTERA

Family	Distinguishing Characteristics	Importance
Arctiidae Tiger moths Fall webworm Saltmarsh caterpillar Fig. 4.6D1, D2	Medium sized, robust, hairy moths with brightly spotted or banded wings held roof-like, and with leading wing edges rounded. Caterpillars usually densely hairy	Caterpillars may cause serious defoliation of trees, shrubs and many agricultural crops
Gelechiidae Gelechiid moths Pink bollworm Angoumois grain moth Fig. 4.6A	Tiny to small usually colorful moths with long, upcurved and pointed labial palps. Larvae pale or pinkish, and hairless	Larvae leafminers, leaf-rollers or leaftyers. One species very de-structive to cottonbolls; another to stored grain
Geometridae Measuring worms Inchworms Cankerworms Fig. 4.6F1, F2	Small light-colored nocturnal moths with body and legs slender, with sparse hairs. Wings usually broad with fine parallel bands. Females of a few species wingless. Larvae with 2 or 3 pairs of prolegs	Larvae destructive to the foliage of trees and shrubs
Gracillariidae Leafblotch miners Fig. 4.6B1, B2	Minute to small variously colored moths with lanceolate wings and with anterior part of body elevated when resting. Larvae flattened with enlarged thorax and rudimentary legs	Larvae make blotch or serpentine mines in the leaves of white oak and other plants
Hesperiidae Skippers	Small to medium-sized, fast erratic-flying butterflies with an-tennae clavate and hooked beyond the club. Wings variously colored and held partly open and raised when at rest. Caterpillars naked and head large and constricted at neck	Larvae mainly attack fo-liage of cereals, grasses, legumes and palms. Usually not destructive

TABLE 4.8 *(Continued)*

Family	Distinguishing Characteristics	Importance
Lasiocampidae Tent caterpillars Lappet moths Fig. 4.6C1, C2	Medium-sized, brown or gray, stout-bodied and hairy moths with pectinate or feathery antennae and front basal part of hind wing expanded and veined. Larvae dark, somewhat hairy with light stripe or keyhole-shaped spots down middle of back	Larvae defoliate trees and shrubs. Some live in webbed nests on the trees
Lycaenidae Gossamer-winged 　butterflies Hairstreaks, 　Blues, Coppers	Small delicate butterflies with metallic-colored wings, sometimes brightly marked; and often with tiny taillike projections on hind wings. Larvae flattened, sluglike	Larvae phytophagus, rarely destructive; a few are predaceous on aphids
Lymantriidae **(Liparidae)** Tussock moths Gypsy moths Brown-tail moths Fig. 4.6E	Medium-sized moths with feathery antennae and leading edges of wings rounded. Wings of females (lacking in some species) white with black marking or vice-versa. Similar to Noctuidae but lack ocelli. Caterpillars hairy, colorful, often with tufts or brushes of hair	Larvae phytophagus: serious pests of forest and shade trees
Noctuidae Cutworm moths Noctuid moths Armyworms Corn earworm Tobacco budworm Fig. 4.6G1, G2	Medium-sized, dull-colored (hind wings sometimes brightly colored), heavy-bodied moths, usually with threadlike antennae and hind wings broader than fore wings. Ocelli present. Caterpillars naked, gray or blackish in color with 5 pairs or sometimes 3 pairs of prolegs	Caterpillars mostly foliage feeders, some fruit-eating or boring. Serious pests of various crops
Nymphalidae Brush-footed 　butterflies Fritillaries Angelwings Admirals Mourningcloak Tortoiseshells	Medium to large-sized butterflies usually with colorful patterned wings and with prothoracic legs much reduced and without claws. Caterpillars naked, usually spiny, often strikingly marked	Caterpillars phytophagus: feed on wide range of plants; sometimes destructive

TABLE 4.8 *(Continued)*

Family	Distinguishing Characteristics	Importance
Olethreutidae Fruit moths Codling moth Oriental fruit moth Fig. 3.19A, B, C	Small moths, brownish to grayish in color. Forewings with bands or mottled areas, somewhat square-tipped; hind wings usually with fringe of long hairs near their base. Larvae whitish, yellowish, greenish or pinkish with dark heads	Larvae very destructive fruit, berry and nut pests: tunnel in the pulp or seeds
Papilionidae Swallowtails	Large colorful, mostly black and yellow butterflies, usually with a taillike projection on the hind wings. Larvae fleshy and naked with an eversible scent gland in the prothorax	Larvae phytophagus: feed on plants in the dill family and on the foliage of various trees and shrubs
Pieridae Whites Sulphurs Orange-tips Fig. 4.7A1, A2	Small to medium-sized butterflies with white orange or yellowish wings with black marginal markings. Front legs well-developed with forked claws. Caterpillars usually green or yellowish, slender, naked or covered with short hair	Larvae are most common on and destructive to the foliage of cruciferous and leguminous crops
Pyralidae Pyralid moths Grass moths Flour moths Fig. 4.7B	Small to medium-sized delicate, usually drab-colored moths; often with labial palps projecting forward snoutlike. Larvae naked with small tubercles bearing setae	Larvae destructive borers in corn, sugarcane, rice and the stems of many other herbaceous plants. Flour moth larvae very destructive to stored cereals and other vegetable products
Sesiidae (Aegeridae) Clearwing moths Fig. 4.7C	Small to medium-sized, brightly colored, slender, day flying moths resembling wasps. Wings mostly transparent almost wholly lacking scales. Forewings long and narrow, hind wings broad. Larvae cylindrical, white or pale with 5 pairs of prolegs	Larvae very destructive borers: tunnel into roots, stems, canes or trunks of herbaceous or woody plants

TABLE 4.8 *(Continued)*

Family	Distinguishing Characteristics	Importance
Sphingidae Sphinx moths Hawk moths Hornworms Fig. 4.7E1, E2	Medium to large, robust moths with spindle-shaped tapered bodies, long narrow forewings and short narrow hind wings; usually with a very long siphoning proboscis. Wings usually brilliantly patterned, some almost clear-winged. Dusk fliers, feed like humming birds. Larvae naked usually with a hornlike structure at dorsal end of abdomen	Larvae phytophagus: very destructive to foliage of solonaceous crops and some trees
Tineidae Clothes moths Fig. 4.7F1, F2	Adults tiny to small tawny or grayish moths with narrow wings, the hind wings tapering smoothly to apex. Siphoning tongue present or absent. Larvae pale with prolegs, some are casebearers	Larvae destructive: feed largely on dried vegetable or animal matter—woolen fabrics silk, cloths, rugs, etc.
Tortricidae Leafroller moths Spruce budworm Fig. 4.7D	Small moths with tan, brown or gray wings, spotted or marbled. When at rest the moths have a bell-shaped appearance. Larvae greenish with scattered hairs arising from small tubercles, when disturbed wriggle violently to escape	Larvae often serious de-foliators, leafrollers or leaftyers of trees and shrubs

Larvae are caterpillars with chewing mouthparts, smooth or hairy bodies, usually with two, three or five pairs of prolegs on the abdomen and with several pairs of ocelli on the sides of the head. Many are serious pests of forest trees and cultivated plants; some infest stored grain, meal and cereals and a few attack various types of fabrics and furs.

Diptera.—Flies, Mosquitoes, Midges (Figs. 4.8, 4.9). This group includes tiny- to medium-sized insects with one pair of wings (rarely none). The second pair are reduced to small stalked knoblike structures called **halteres.** The compound eyes are very large; the legs are relatively long; and the antennae are usually short and variously shaped (frequently **aristate**). The mouthparts may be sponging, piercing-

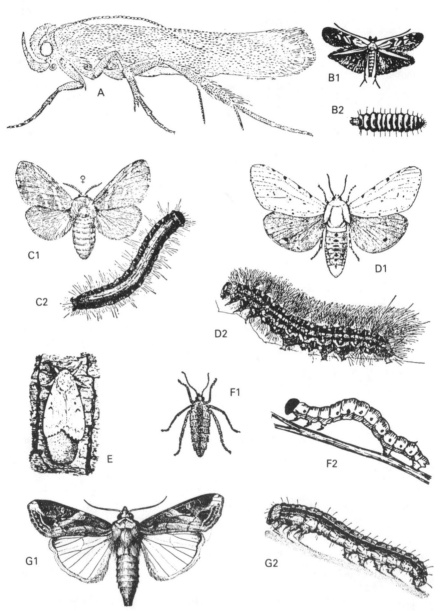

(A, from History of Entomology; B, after Comstock; C, D, G, courtesy U.S.D.A.; E, courtesy Connecticut Agricultural Experiment Station; F, after U.S.D.A.)

FIG. 4.6. FAMILIES OF LEPODOPTERA

A, **Gelechiidae,** potato tuber moth; B1, B2, **Gracillariidae,** solitary oak leafminer and larva; C1, C2, **Lasiocampidae,** eastern tent caterpillar and larva; D1, D2, **Arctiidae,** salt-marsh caterpillar; E, **Lymantridae,** gypsy moth depositing egg mass; F1, F2, **Geometridae,** spring canker worm and larva; G1, G2, **Noctuidae,** fall armyworm moth and larva.

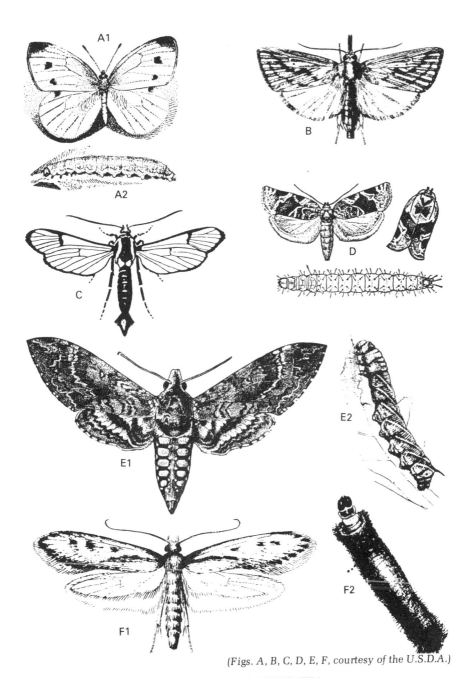

(Figs. A, B, C, D, E, F, courtesy of the U.S.D.A.)

FIG. 4.7. FAMILIES OF LEPIDOPTERA

A1, A2, **Pieridae,** imported cabbageworm butterfly and larva; B, **Pyralidae,** sugarcane borer moth; C, **Sesiidae,** peachtree borer moth; D, **Tortricidae,** red-banded leafroller moth and larva;E1, E2, **Sphingidae,** tobacco hornworm moth and larva; F1, F2, **Tineidae,** casemaking clothes moth and larva.

TABLE 4.9

THE ECONOMICALLY IMPORTANT FAMILIES OF DIPTERA

Family	Distinguishing Characteristics	Importance
Anthomyiidae Rootmaggot flies Fig. 4.8D	Small dark-bodied flies, similar to house flies, but the wings are finely hairy and held flat and parallel to the body. First posterior cell in wing wide open, anal vein extends to margin	Many larvae (maggots) very destructive to roots of garden crops. A few species are leafminers
Calliphoridae Blow flies Blue-bottle flies Screw-worm flies Fig. 4.8A	Small to large robust flies, generally with a metallic blue or green abdomen and arista of antennae plumose to tip	Larvae (maggots) breed in decaying flesh and manure. Screw-worm fly maggots may infest living tissues (myiasis)
Cecidomyiidae Gall midges Gall gnats Hessian fly Fig. 4.8B1, B2	Minute, delicate, slim-bodied flies with long legs, fairly long antennae, and with reduced wing venation. Larvae—tiny maggots sometimes brightly colored	Maggots may produce galls on plants or feed on plants
Ceratopogonidae Biting midges Punkies No-see-ums Fig. 4.8G	Tiny blood-sucking flies with narrow naked or hairy wings held flat and with few veins. Tibiae with spurs. Larvae long, slim semi-aquatic or aquatic	Pestiferous, painful bites. Suck blood—attack humans, animals and many insect species
Chironomidae Midges Fig. 4.8C1, C2	Small mosquito-like insects without beak and no scales on wings. Larvae called blood-worms, aquatic scavengers	Food for many fresh-water fish—part of aquatic food chains
Culcidae Mosquitoes Fig. 4.8F1, F2	Small, fragile, elongate, long legged insects—with scales or dense hairs on veins of wings, with long beak containing piercing-sucking mouthparts, and with weakly plumose antennae. Larvae (wrigglers) aquatic	Adults (females) feed on blood of humans and animals. Vectors of yellow fever, malaria, encephalitis, dengue and filariasis

TABLE 4.9 *(Continued)*

Family	Distinguishing Characteristics	Importance
Hippoboscidae Louse flies Fig. 4.8E	Small leathery, flattened, winged or wingless insects with head partially enveloped by thorax, with piercing-sucking mouthparts and with indistinctly segmented rounded abdomen	Blood-sucking ectoparasites on birds. One species (sheep ked) found on sheep
Muscidae House flies Stable flies Face flies Tsetse fly Fig. 4.9A, B	Small to medium-sized robust grayish flies with aristate antennae, sponging or piercing-sucking mouthparts and with wings usually held flat away from abdomen when at rest. Larvae (maggots) breed in manure and decaying plant or animal matter	Some feed on blood of hosts, others just annoying and unsanitary. Many are vectors of serious human and animal diseases
Oestridae Bot flies Warble flies Fig. 4.9D	Large dark brown; robust, densely hairy flies resembling bees; with vestigial mouthparts. Swift flyers. Larvae—robust maggots called bots	Larvae endoparasitic in wild and domestic animals, rarely in humans. May seriously affect health
Psychodidae Moth flies Sand flies Fig. 4.9C	Minute to small hairy mothlike flies with hairy wings held rooflike over the body. Phlebotomus species blood-sucking	Most northern species infest drains and sewers —harmless. Sand flies vectors of several tropical diseases
Sarcophagidae Flesh flies Fig. 4.9E	Medium-sized flies with thorax typically gray with three black stripes, and with a checkered gray and black abdomen (not metallic or bristly)	Adults harmless. Maggots feed on decaying flesh and animal matter. Some are parasitoids of other insects
Simulidae Black flies Buffalo gnats Fig. 4.9F	Small dark-colored, humpbacked flies with broad wings, short legs and blood-sucking mouthparts. Larvae aquatic, live in streams	Females vicious bloodsuckers of humans and animals. Very annoying, sometimes deadly pests in the spring in cool regions. Filarial disease vectors in tropical regions

TABLE 4.9 *(Continued)*

Family	Distinguishing Characteristics	Importance
Syrphidae Flower, Syrphid Hover and Drone flies Fig. 4.9G	Small to large brightly-colored flies resembling bees or wasps, but do not bite or sting. Common in the fall around chrysanthemums and other flowers	Many larvae predaceous on aphids—beneficial; others breed in decaying vegetation or in polluted water (rattailed maggots)
Tabanidae Horse flies Deer flies Fig. 4.9H	Medium-sized to large rather robust flies with large often irridescent or brightly colored eyes, and with clear or clouded wings. Larvae: aquatic and predaceous	Females blood-sucking, persistent pests of live-stock and humans. Some vectors of serious human diseases
Tachinidae Tachina flies Fig. 4.9I	Medium-sized to large bristly flies with a prominent postscutel-lum (ridge or bulge just behind the scutellum) and usually with bare arista	Larvae valuable parasitoids of harmful insects
Tephritidae Fruit flies Fig. 4.9J	Small to medium-sized flies with spotted, banded or mottled wings slowly moved up and down when at rest. Larvae are typically maggot-like	Very destructive pests of fruits—adults deposit eggs in fruits, maggots tunnel in the tissues. A few are leafminers

sucking or a combination of both. Many species feed on the blood of animals and humans; some on the nectar of flowers and others are omnivorous scavengers.

The larvae (maggots) are light-colored, wormlike and legless, with reduced pointed heads and hooklike mouthparts in many species. Some species are aquatic; those that are terrestrial feed in plant tissues, in decaying plant or animal matter, or within the living bodies of humans and animals. Some are destructive pests of cultivated plants. This order also includes vectors of many serious diseases of humans and animals; some predators or parasites of other insects; and some helpful pollinators.

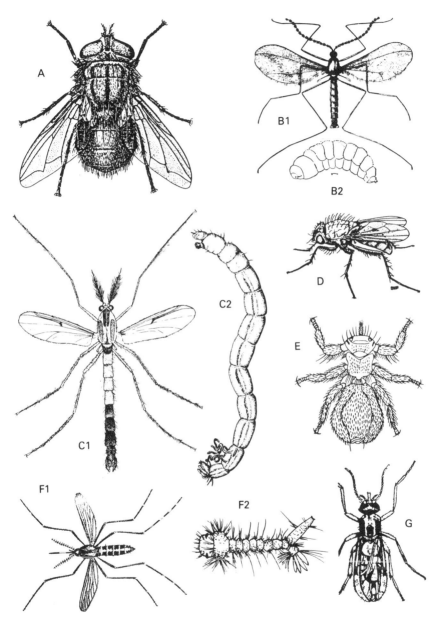

(Figs. A, B, E, courtesy U.S.D.A.; C, courtesy Illinois Natural History Survey;
D, courtesy North Carolina Agricultural Experiment Station; F1, courtesy
New Jersey Agricultural Experiment Station; G, after Essig, 1942)

FIG. 4.8. FAMILIES OF DIPTERA

A, **Calliphoridae**, screw-worm fly; B1, B2, **Cecidomyiidae**, Hessian fly; C1, C2,
Chironomidae, male midge and larva; D, **Anthomyiidae**, cabbage maggot fly; E, **Hippoboscidae**, sheep ked; F1, F2, **Culicidae**, salt marsh mosquito and larva; G,
Ceratopogonidae, the little gray punkie.

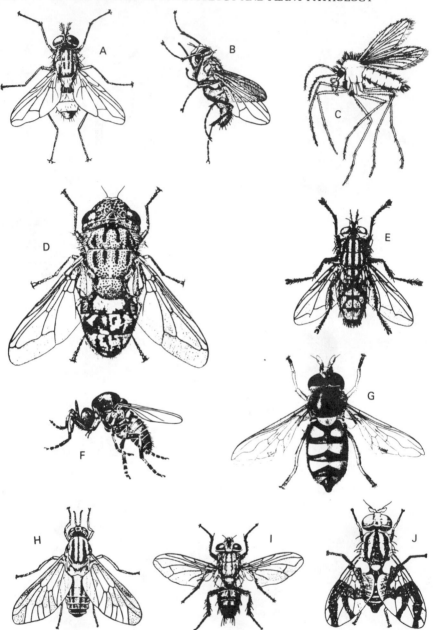

(Figs. A, B, D, E, F and I, courtesy U.S.D.A.; C, courtesy U.S. National Science
Foundation; G, courtesy Maine Agricultural Experiment Station;
H, courtesy Arkansas Agricultural Experiment Station; J, after Snodgrass)

FIG. 4.9. FAMILIES OF DIPTERA

A, **Muscidae,** housefly; B, **Muscidae,** stablefly; C, **Psychodidae,** pappataci sandfly; D,
Oestridae, sheep botfly; E, **Sarcophagidae,** redtailed flesh fly; F, **Simulidae,** black fly;
G, **Syrphidae,** syrphid fly; H, **Tabanidae,** deerfly; I, **Tachinidae,** tachinid fly; J,
Tephritidae, apple maggot fly.

Hymenoptera.—Sawflies, Bees, Wasps, Ants, Chalcids (Fig. 4.10). The order consists of minute to large insects usually with two pairs of sparsely-veined membraneous wings of unequal size; the forewings are larger and longer, frequently with a small heavily pigmented area (the **stigma**) in the upper edge of the wing near the apex. The mouthparts are chewing but in most bees parts are modified for lapping up liquids as well. The base of the abdomen is constricted in all species except the sawflies and horntails. The females usually have prominent ovipositors or stings. Many species are social with caste differences.

Larvae may be caterpillarlike, maggotlike or grublike. When caterpillarlike, they have well developed chewing mouthparts, one distinct occellus on each side of the head and six to eight pairs of prolegs; when maggotlike or grublike they are legless and have poorly developed heads with mouthparts chewing or reduced. The larvae of sawflies feed on plants; those of bees, wasps, and ants are fed and cared for by a worker caste or the adult females; others are parasitoids on insects; and some are scavengers. Except for the sawflies, wasps and ants, the hymenoptera are a most valuable group of insects involved in the pollination of crops, the production of honey and wax and the parasitism of other insects. Bees, wasps and some ants may inflict painful and sometimes fatal stings to humans and animals.

TABLE 4.10

THE ECONOMICALLY IMPORTANT FAMILIES OF HYMENOPTERA

Family	Distinguishing Characteristics	Importance
Apidae Honeybees Bumblebees Fig. 4.10A, B	Social or solitary, medium-sized to large, robust or elongate-oval, hairy-bodied insects; hind legs usually adapted for pollen collecting. Generally seen around flowers. Young grublike, helpless, fed by workers or adult females	Valuable plant pollinators. Produce honey and wax
Andrenidae **Halictidae** Flower bees Andrenid bees Fig. 4.10C	The two families very much alike—solitary or colonial, small- to medium-sized, usually metallic colored. With two sutures below each antennal socket (Andrenidae), lacking two sutures (Halictidae)	Both families include valuable plant pollinators

TABLE 4.10 *(Continued)*

Family	Distinguishing Characteristics	Importance
Braconidae Branconid wasps Fig. 4.10D	Relatively small wasplike insects with slim, sparsely-veined wings and relatively long many-segmented antennae. Females with long or short ovipositors	Valuable parasitoids of the larvae of many insect pests
Cephidae Stem sawflies Fig. 4.10E	Small slender elongate sawflies with abdomen laterally compressed	Larvae bore in the stems of grasses, grains and into berries
Chalcidae Chalcids Fig. 4.10F	Minute to small compact insects with two pairs of almost veinless wings generally held flat over body; with short elbowed antennae. Many metallic in color	Endo-parasitoids on the eggs or larvae of many crop pests
Formicidae Ants Fig. 4.10H	Tiny- to medium-sized insects, winged or wingless with geniculate antennae and constricted waists; with chewing mouthparts; some with stings	Biting or stinging pests in buildings or outdoors; harbor aphids, feed on honeydew and plant or animal products. Some built mounds in fields
Ichneumonidae Ichneumons Fig. 4.10K	Small to large slender wasplike insects with long filiform antennae, long needlelike ovipositors and elongate wings with well-defined venation	Most are internal parasitoids of the immature stages of many insects including wood-borers
Sphecidae Mud daubers Thread-waisted wasps Digger wasps Fig. 4.10I	Solitary, usually slender metallic-colored wasps of medium size, usually with stalked abdomens and pronotum terminating laterally in a rounded lobe	Provision nests with caterpillars, spiders, or adult insects. Stings painful
Tenthredinidae Sawflies Fig. 4.10J	Usually small- to medium-sized, robust, broad-waisted insects with short, usually filiform, antennae and front tibiae with two apical spurs. Larvae: caterpillar-like with 6-8 pairs of prolegs; a few sluglike	Larvae destructive—consume the foliage of trees and shrubs; some are leafminers

TABLE 4.10 *(Continued)*

Family	Distinguishing Characteristics	Importance
Vespidae Social wasps Potter wasps Paper wasps Yellowjackets Hornets Fig. 4.10G	Medium-sized to large social or solitary insects, robust to elongate-slender with constricted waists. Wings usually folded lengthwise at rest. Nests in the open or in cavities) in the ground	Adults provision nests with paralyzed insects or feed their young during growth. Stings painful. Giant hornet debarks lilac
Xylocopidae Carpenter bees	Medium to large, robust hairy bees, metallic-colored somewhat flattened abdomens. Resemble bumblebees but top of abdomen hairless	Cause structural damage—make round nesting holes in wood or plant stems

Neuroptera.—Lacewings, Alderflies, Dobsonflies, Aphid Lions. This group consists of small to large slender insects with large, widely separated compound eyes, long thread-like many-segmented antennae, chewing mouthparts, and with two pairs of similar many-veined membraneous wings, usually held slanted over the body when at rest. Larvae may be either aquatic (alderflies, dobsonflies) or terrestrial (lacewings, aphid lions) often with long sickle-like jaws modified for sucking up the body fluids of insect hosts. Neuroptera are of some importance in aquatic food chains. The lacewings in both the adult and larval stages prey on aphids and other small insects and are beneficial.

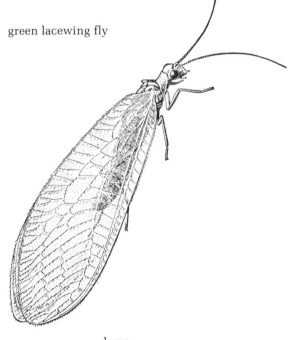

green lacewing fly

larva

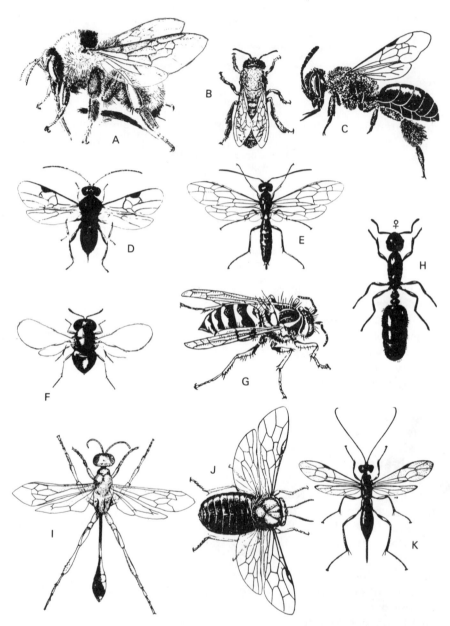

Fig. A, after C.O. Mohr; B, E, F, H, J, courtesy U.S.D.A.; C, after Woodsworth from Insects of Western North America; D, courtesy Entomology Society of America; G, courtesy Illinois Natural History Survey; I, courtesy Utah Agricultural Experiment Station; K, courtesy of Roher)

FIG. 4.10. FAMILIES OF HYMENOPTERA

A, **Apidae,** yellow bumblebee; B, **Apidae,** honeybee; C, **Andrenidae,** burrowing bee; D, **Braconidae,** braconid parasite of European cornborer; E, **Cephidae,** wheat stem sawfly; F, **Chalcidae,** chalcid gypsy moth parasite; G, **Vespidae,** yellow jacket; H, **Formicidae,** little black ant worker; I, **Sphecidae,** digger wasp; J, **Tenthredinidae,** red-headed pine sawfly; K, **Ichneumonidae,** ichneumonid parasite.

Tricoptera.—Caddiceflies. These are small- to medium-sized insects with poorly developed chewing mouthparts; long slender antennae; and with four hairy membraneous wings, with few cross veins, held rooflike over the body. The larvae are aquatic and caterpillarlike plant feeders or predaceous with filamentous abdominal gills. Some are casemaking plant feeders using bits of leaves, twigs, sand grains or other material to form cases characteristic of the species; others are predaceous net-making or free-living. The larvae and adults are economically of value as a part of aquatic food chains.

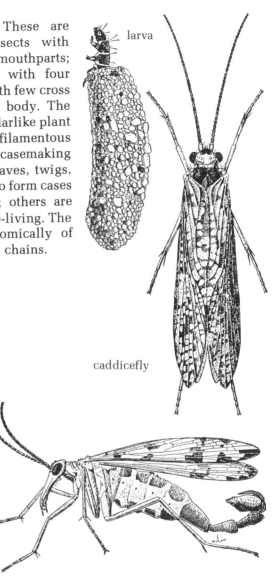

larva

caddicefly

Mecoptera.—Scorpion-flies. Scorpionflies are medium-sized slender omnivorous insects with chewing mouthparts located at the end of a ventrally prolonged snoutlike head; with two pairs of long narrow membraneous wings similar in size, shape and venation; with large compound eyes; and with long slender many-segmented antennae. Their caterpillarlike larvae live either underground or on the surface feeding as predators or as scavengers on decaying organic matter. The males in the family Panorpidae have scorpionlike tails. The group is of little economic importance.

Siphonaptera.—Fleas. These small brownish wingless ectoparasites of humans, animals and birds are characterized by laterally flattened bodies covered with numerous backward-projecting spines and bristles; often with some on the head or thorax enlarged to form comblike rows (**ctenidia**). Eyes may be present or absent; antennae are short and folded into grooves in the head; legs are fitted for jumping with greatly enlarged coxae; and the mouthparts, with long palpi, are piercing-sucking and used in feeding on the blood of its hosts. The larvae are whitish, elongate-slim, legless, with the body sparsely covered with bristly hairs and with a fairly distinct head with chewing mouthparts. The larvae are usually found in the litter of the resting places of their hosts where they feed on organic debris and the feces of the adult fleas. Fleas are highly irritating to their hosts but more importantly may be vectors of bubonic plague, endemic typhus and tapeworms to humans. One species, the chigoe flea, found in warm climates, burrows into the skin (usually around the toes) of humans and animals and produces ulcerlike sores.

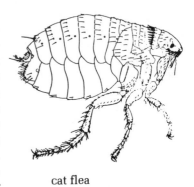

cat flea

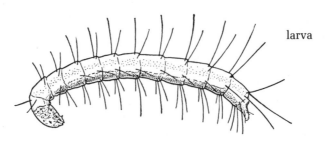

larva

QUESTIONS FOR DISCUSSION

1. Why is the classification of insects important to people that deal with insect control?
2. What would happen if there were no standard system of classification?
3. Why are not definitions of a species foolproof?
4. How do varieties, races, or strains originate?
5. What is the procedure in naming a new, formerly unclassified insect or other organism?
6. Why is it impossible for a person to be proficient in classifying all types of insects?
7. What are the characteristics common to all insects?
8. Is there any value in being able to identify insects to the correct order? To the correct family?
9. Which orders contain species most harmful to plants? To animals?
10. Give an outstanding characteristic of each order that includes important plant pests.
11. Cite examples of beneficial insects.
12. What are the harmful groups, as far as crops are concerned, in the animal kingdom? Give examples.
13. What features distinguish insects from all other organisms?
14. Why are nematode worms usually classified with disease-producing organisms?
15. Can you advance any theories to explain why protozoa are unimportant as the cause of plant diseases but of great importance in the cause of animal and human diseases?

Chapter 5

Harmful Invertebrates Other Than Insects

ARTHROPODS OTHER THAN INSECTS

The Symphylans (Symphyla) (Fig. 5.7A)

Symphylans are tiny, active slender arthropods, white in color; with chewing mouthparts, 14–22 body segments and 10–12 pairs of legs. The head is well developed with long, slender many-segmented antennae. Development occurs with very little metamorphosis—when hatched they possess only six pairs of legs and short antennae but with each molt (six) they acquire an additional pair of legs and longer antennae. For the most part symphylans are subterranean. One species, the garden symphylan, is a serious pest of many crops in fields, gardens, seedbeds and greenhouses injuring germinating seeds, seedlings and rootlets.

The Millipedes (Diplopoda)

Some 1000 species of this class have been described, but most of them feed on manure and decaying vegetable matter on or in the soil. Some species (Fig. 5.1), however, may attack the stems and underground parts of plants, tunnel into the root vegetables, eat planted seed, and bore into fruits that are in contact with the ground. Most millipedes have an offensive odor that tends to protect them from their predators. They have chewing mouth parts and no metamorphosis. The young, when they hatch from eggs laid in clusters in the soil, resemble the full-grown millipede except for size, number of body segments, and legs. Like insects, they breathe through spiracles and undergo a series of molts as they grow. When at rest or disturbed, they generally curl up like a watch spring. They grow very slowly, and in northern regions there is probably only one life cycle a year. Millipedes overwinter in partly to fully grown stages in protected places in the soil or under objects resting on the soil.

Centipedes (Chilopoda) and sowbugs (Crustacea) are usually of little economic importance. (See Table 4.2 and Fig. 5.1.)

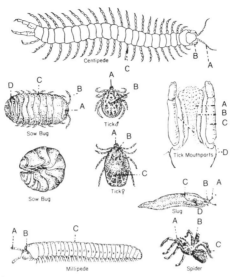

FIG. 5.1 OTHER ARTHROPODS AND A SLUG (MOLLUSCA)

Centipede-A, Antenna; B, Poison-fang; C, Body; *Sowbug*-A, Head; B, Antenna; C, Thorax; D, Abdomen; *Tick*-A, Mouthparts; B, Scutum; C, Abdomen; *Tick mouthparts*-A, Hypostome; B, Chelicera; C, Pedipalp; D, Capitulum; *Slug*-A, Tenticles; B, Stalked eyes; C, Mantle; D, Respiratory pore; *Millipede*-A, Antenna; B, Head; C, Body; *Spider*-A, Palps; B, Cephalothorax; C, Abdomen.

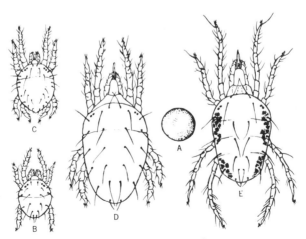

(Redrawn from Peairs,
Insect Pests of Farm, Garden, and Orchard)

FIG. 5.2. THE DEVELOPMENTAL STAGES OF THE TWO-SPOTTED MITE
A, Egg; B, Larva; C, Protonymph; D, Deutonymph; E, Adult.

The Spiders (Araneida) and the Mites and Ticks (Acarina or Acari)

With the exception of a few poisonous ones the spiders are beneficial, feeding mainly on insects and on other small organisms. The ticks (Figs. 5.1, 5.7F) are pests of man and animals, sucking their blood and transmitting many serious diseases. The mites (Fig. 5.2) include serious pests of both animals and plants. Spiders have a head and thorax fused together to form a *cephalothorax* from which the four pairs of legs originate and an abdomen that is constricted at the point of attachment to the cephalothorax (Figs. 5.1, 5.7D). Mites and ticks are distinguished from spiders in that the abdomen of mites and ticks is joined to the cephalothorax with little or no indication of a division between these two regions, the entire body tending to present a sac-like appearance. Mites and ticks differ mainly in size—the smaller species that are hardly visible to the eye are referred to as mites or red spiders.

General Characteristics of Spider Mites and Red Spiders.—Mites and red spiders are common and destructive pests of many different plant species grown under glass or outdoors. They pierce the plant tissue with their sharp mouthparts, sucking up the sap and destroying the chlorophyll (Fig. 5.3). This injury on foliage may result in a gray or brown mottling of the foliage or in the formation of galls or growths. Many species of mites are predatory or parasitic on insects and may be considered beneficial.

Mites have one or more pairs of simple eyes placed laterally on the forward part of the body. They breathe either by tubular tracheae or directly through the body surface. The spiracles (*stigmata*) are generally located on the dorsal side near the mouth parts. Most species are oviparous, some are ovoviviparous, and a few are viviparous. Most of the mites that attack plants are divided into two groups: mites or red spiders (Tetranychidae) and gall, rust, or blister mites (Eriophyidae). Since these two groups differ greatly in habits they are considered separately.

The Red Spiders or Spider Mites (Tetranychidae).—These are actively crawling mites in which the eyes are recognizable as red spots on the front end of the body, the ends of the tarsi have peculiar tenent hairs each with a transverse hook at the tip, and the mouth parts are needle-like and covered at the base by a mandibular plate (Fig. 5.3). The usual mode of respiration is by means of a simple tracheal system with stigmata dorsal above the mandibular plate. Mites commonly cover leaf surfaces with fine webbing.

The life stages of these mites consist of egg, larva (with 3 pairs of legs instead of 4), 2 nymphal stages (protonymph and deutonymph) that resemble the adult, and the adult (Fig. 5.2). Each molt is preceded by a quiescent stage of a day or two during which the mites are very resistant to chemical control.

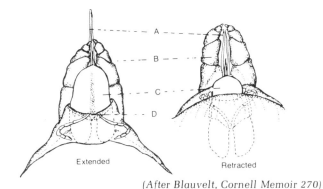

(After Blauvelt, Cornell Memoir 270)

FIG. 5.3. DORSAL VIEW OF THE MOUTHPARTS OF THE TWO-SPOTTED MITE (ENLARGED)

A, Chelicera; B, Palps; C, Mandibular plate; D, Stigmata.

The females lay a few eggs each day for about 2–3 weeks. Since the complete life cycle takes from 15–20 days under favorable conditions, the broods overlap, and all stages may be found at the same time. There are several generations a year for most species. Some species overwinter in the egg stage on their host plants, whereas others overwinter as adults and nymphs under the bark of trees, under dead leaves, or in the ground. In southern states mites may breed continuously, with their development slowed up considerably during the colder months.

Foliage injured by mites first takes on a grayish, speckled appearance and later turns rusty or brownish (Fig. 5.4). Plants or their fruits are stunted, and in severe cases defoliation occurs. A continuous, heavy infestation usually results in plants becoming greatly stunted or killed outright. Mites prefer the undersides of foliage but may be found on both sides under conditions of heavy infestation. They flourish during warm, dry weather and are not generally troublesome during cool, wet seasons. The seasonal life histories of two common representative mites, the European red mite and the two-spotted mite, are illustrated in Fig. 5.5.

The Gall, Rust, and Blister Mites (Eriophyidae).—The mites in this group (Fig. 5.6) are extremely minute, with various species being called gall mites, russet mites, rust mites, or blister mites from the nature of their injury to foliage or fruit. Some mites in this group cause abnormal development of the epidermal plants cells or a deformation of the plant hairs which then appear as brightly colored patches on the foliage and are often mistaken for fungous growth. This type of abnormal growth is termed an *erineum*.

The mouth parts of the Eriophyidae are similar to those of the Tetranychidae. The mites use these mouthparts in penetrating leaf tissue which produces galls or blisters; in feeding externally which produces

FIG. 5.4. THE TWO-SPOTTED MITE, ITS EGGS, AND INJURY ON A SWEET PEA
LEAF

silvery or rusty splotches on foliage or fruit; or in irritating the tissue
which produces an erineum. The mites have no external metamorphosis, both the nymphs and the adults having only 2 pairs of 5-
segmented legs with a feathered hair extending from the tip of each
tarsus. The life cycle generally consists of the egg, two nymphal stages,
and the adult. The tomato russet mite has a rather unusual type of

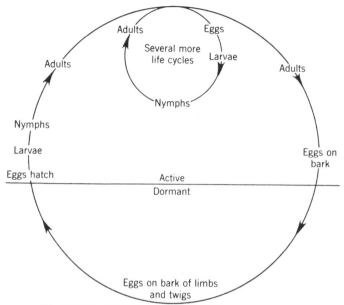

FIG. 5.5. THE LIFE HISTORY OF THE EUROPEAN RED MITE
The life history of the European red mite *(Panonychus ulmi)*. The life history of the two-spotted mite *(Tetranychus urtlcae)* differs only in that it overwinters in the adult stage under bark, among leaves, or in the soil.

reproduction. The unfertilized females produce males which in turn fertilize the females to produce female progeny. The Eriophyidae are whitish and have a somewhat worm-like appearance with a dorsal thoracic shield and a tapering, ringed abdomen with a pair of caudal setae extending from the end. The life cycle of these mites varies from 1–3 weeks under normal conditions.

Mites and Ticks of Medical Importance.—Several families of mites are serious ectoparasites of humans, mammals and birds. The **Acaridae** (*Tyroglyphidae*) are minute whitish mites that infest grain, flour, meal, dried fruits and meats. They are the cause of a severe dermatitis on people handling such stored products. They bear such names as grain mites, flour mites, cheese mites, mushroom mites and itch mites.

The **Sarcoptidae** (Fig. 5.7B). These mites are very minute, whitish hemispherical with 8 legs that do not extend beyond the body margin. They are commonly known as sarcoptic itch mites or scabies mites. They burrow in the epidermis of the skin of humans and mammals, producing inflamed, itching scabby areas referred to as scabies or mange.

The **Psoroptidae** (Fig. 5.7C). These are also very tiny whitish mites with oval bodies, 8 legs extending beyond the body margin and bearing tarsal suckers on long, segmented stalks. They tend to limit their attacks to definite parts of the mammals' body such as the feet, tail, neck or ears. The psorptic mange mites do not burrow in the skin but settle at the base of the hairs of the host, pierce the skin causing inflammation, exudations, and crusty scabs.

TABLE 5.1

SOME IMPORTANT PLANT MITES

Name	Host Plants
Tetranychidae	
Boxwood mite (Eurytetranychus buxi)	Boxwood
Citrus red mite (Paratetranychus citri)	Citrus
Clover mite (Byrobia praetiosa)	Deciduous fruit trees, clover, other forage crops
European red mite (Panonychus ulmi)	Deciduous fruit and other trees
Pacific mite (Tetranychus pacificus)	Fruit trees, milkweed
Six-spotted mite (Tetranychus sexmaculatus)	Citrus
Southern red mite (Oligonychus ilicis)	Deciduous plants
Spruce mite (Oligonychus ununguis)	Needled evergreens
Two-spotted mite (Tetranychus urticae)	Deciduous plants
Eriophyidae	
Citrus rust mite (Phyllocoptruta oleivora)	Citrus
Maple bladdergall mite (Vasates quadripedes)	Red and sugar maples
Pear leaf blister mite (Eriophyes pyri)	Pear and apple
Apple rust mite (Aculus schlechtendali)	Apple, pear, prune in Northwest
Tomato russet mite (Vasates lycopersici)	Potato, tomato, petunia, night-shade
Others	
Bulb mite (Rhizoglyphus echinopus)	Bulbs
Cyclamen mite (Steneotarsonemus pallidus)	Cyclamen, begonia, petunia, dephinium, snapdragon, etc.
Grain mite (Acarus siro)	Mushrooms, cheese, grains, flour, cereals

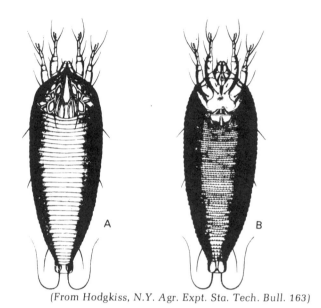

(From Hodgkiss, N.Y. Agr. Expt. Sta. Tech. Bull. 163)

FIG. 5.6. THE MAPLES GALL MITE *(VASATES QUADRIPEDES)*
A, Dorsal view; B, Ventral view.

The **Demodicidae.** The follicle mites are extremely minute with elongated, transversely striated abdomens and 4 pairs of very stubby legs located at the anterior end of the body. They are usually found in the hair follicles and sebaceous glands around the nose, eyelids and head of humans and mammals causing dermititis or cystic swellings.

The **Trombiculidae** (Fig. 5.7E). The minute chigger mites are parasitic on humans and animals only in the 6-legged or larval stage. The nymphs and adults are thought to be predatory on eggs of insects of other small organisms. Chiggers do not feed on blood but inject digestive juices in the epidermal cells and feed on the predigestive products. This results in an intense itching and a severe dermititis. Chiggers are usually abundant in grassy tropical or subtropical areas. Two species found in western Pacific regions are known vectors of a serious rickettsial disease of humans, scrub typhus.

The **Hard-bodied Ticks** (Ixodidae) (Fig. 5.1). The ticks in this family are small- to medium-sized bloodsucking parasites of humans and animals with leathery rounded to oval bodies and a hypostome with re-

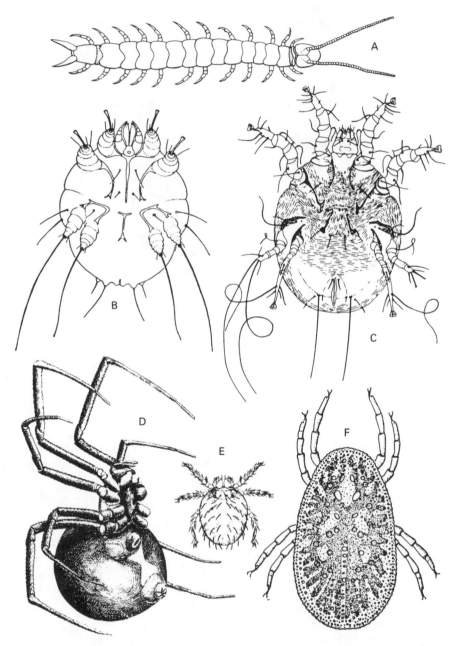

(Figs. A, E, F, courtesy U.S.D.A.; B, after Hermes; C, courtesy National Pest Control Association; D, courtesy Utah Agricultural Experiment Station)

FIG. 5.7. SYMPHILA AND MEDICALLY IMPORTANT FAMILIES OF ACARINA

A, **Symphila,** garden symphilan; B, **Sarcoptidae,** hog itch mite; C, **Psoroptidae,** sheep scab mite; D, **Araneida,** black widow spider; E, **Trombiculidae,** scrub typhus chigger mite, larval stage; F, **Argasidae,** fowl tick.

curved teeth. The first instar ticks (seed ticks) have only three pairs of legs and only acquire a fourth pair after molting. A *scutum* (dorsal shield) is present and covers the whole body in the male but only part of the body in females. They remain attached to their host by their mouthparts while feeding and only drop off when fully engorged some days later.

The **Soft-bodied Ticks** (Argasidae) (Fig. 5.7F). These differ from the hard-bodied ticks by the absence of a scutum, a more or less roughened or wrinkled integument, and a capitulum concealed from above on the anterior ventral side of the body. They are generally nocturnal intermittent feeders in all the stages beyond the larval or seed tick stage, hiding in cracks and crevices during the day.

Both the Ixodid and Argasid ticks are vectors of many serious Rickettsial, bacterial and viral diseases of humans and animals.

THE SNAILS AND SLUGS (MOLLUSCA)

The phylum Mollusca, class Gastropoda, the slugs and snails (Fig. 5.1) may be harmful to plants, especially in storage cellars, greenhouses, mushroom beds, and vegetable gardens. Snails have hard, calcareous, spiral shells of various forms enclosing their bodies, whereas slugs lack such a protecting shell. A sticky, mucous secretion exudes from the bodies of these gastropods, which leaves a trail along the ground and on the plants where they crawl.

These creatures possess a soft, shield-like covering on the fore part of the back and sides, called a mantle cavity or "lung." Air enters this "lung" through a small hole in the right side of the mantle. The head is well defined and bears a pair of tentacles and a pair of stalked eyes that may be extended or withdrawn. For mouth parts, they have a peculiar, tooth-bearing ribbon (called a *radula*) in the mouth cavity, which is used as a rasping organ in feeding. With these mouth parts, they gnaw large, irregular holes in foliage or in plant parts. Unlike insects, the gastropods have a chambered heart and a system of closed vessels with special capillaries extending into the respiratory organ or "lung."

The snails and slugs are hermaphroditic (both sex organs in the same individual) but not self-fertilizing. They are mainly oviparous, with little metamorphosis taking place during their development. Their spherical, glistening eggs are deposited in masses under objects in damp places throughout the active season so that all stages of snails and slugs may be found at the same time. It takes at least a year for most species to reach maturity.

Slugs are nocturnal and come out only after dark to feed. In the daytime they may be found in dark, damp places along hedgerows, in soil cracks, and underneath damp refuse. In greenhouses they are commonly found under pots and flats. They pass the winter below the frost line in the ground, in drain pipes, in cellars, in greenhouses, in pits, on well walls, under mulches, and along foundations. They evidently cannot withstand temperatures much below freezing. Some species of slugs when full grown may attain a length of nearly a foot, although most species are not more than a few inches long. They tend to be most troublesome in cool, damp seasons or in places kept damp by poor drainage or by irrigation.

Few species of land snails are harmful to cultivated plants. Some species have been observed feeding on citrus foliage and other plants in the tropics and in greenhouses. Thirty-two species of garden slugs have been reported in the United States. The native species are relatively unimportant as crop pests, but occasional outbreaks occur. Three introduced species, the spotted garden slug, *Limax maximus*, the tawny garden slug, *Limax flavus*, and the true garden slug, *Deproceras agrestis*, are the main pests of gardens and greenhouse plants in the seaboard states. The true garden slug is the smallest of the three species and also the most common and most destructive.

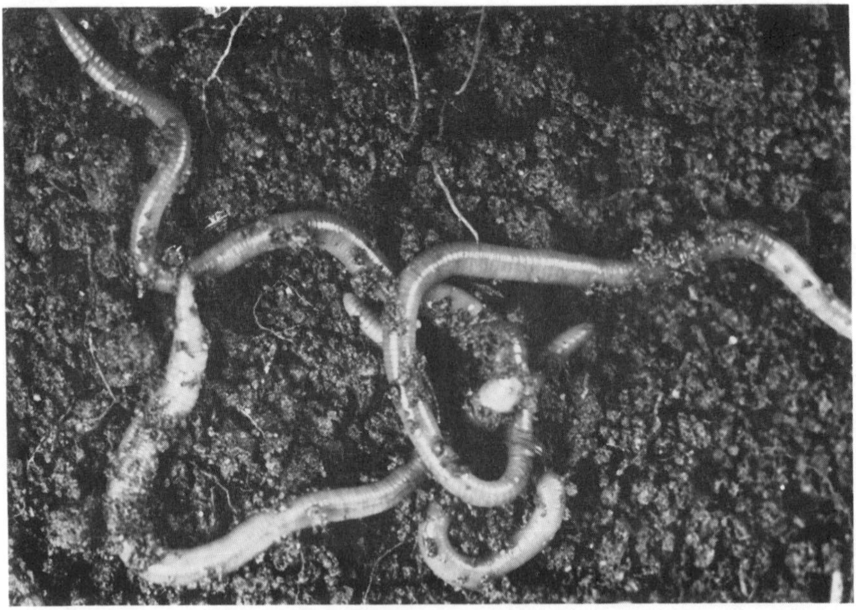

FIG. 5.8. EARTHWORMS INDICATE A RICH SOIL

THE EARTHWORMS (ANNELIDA)

The earthworms are beneficial to plants because they help break down organic matter, aerate the soil, and mix soil layers. On golf greens they are considered harmful because of their casts (excreta), left on top of the close-cut greens. They should be encouraged in fields and gardens by adding organic matter to the soil in the form of animal manures and composted plant remains. See Fig. 5.8.

QUESTIONS FOR DISCUSSION

1. How do millipedes differ from insects?
2. Where are millipedes usually found?
3. Do millipedes undergo a metamorphosis before reaching the adult stage?
4. Distinguish between millipedes and centipedes.
5. In what ways do mites differ from insects?
6. Describe the metamorphosis of tetranychid mites.
7. How do tetranychid mites differ from eriophyid mites?
8. Since mites are wingless, how are they able to disseminate themselves so well?
9. What makes mites difficult to control?
10. How do slugs differ from the other pests considered so far?
11. What environmental conditions favor the development of slugs?
12. Where are slugs generally found?

Chapter 6

The Nematode Worms (phylum Nemata)

Nematode worms closely parallel insects in abundance, distribution, and adaptability. They have been found wherever life exists—in desert areas, ocean waters, fresh waters and soils from the arctics to the tropics. Some are serious internal parasites of vertebrates and invertebrates, and bear such names as hookworms, roundworms, whipworms, pinworms, gapeworms, filarial worms, and trichina worms. Some parasitize such cold-blooded forms as arthropods and mollusks. Others feed on bacteria, fungi, algae, protozoa, and their own kind. But many species that inhabit soils parasitize plants and may produce serious crop losses. Practically every crop or plant species has one or more nematode species as a host. Many plant diseases formerly attributed to fungi, viruses, or soil deficiencies are now known to be caused by nematodes.

Nematode Characteristics

Most plant parasitic nematodes are so small that they cannot be seen without the aid of a microscope. Many thousands may be present around or in the roots of a plant and yet remain undetected without the proper diagnostic procedures. When isolated in water under a microscope and magnified some 100 times, they look like minute, sluggish, or active undulating worms, white in reflected light, transparent in transmitted light. Most are thread-like and range in length from 0.3–2.0 mm or more. Higher magnifications with transmitted light reveal the internal organs through an unsegmented but annulated body wall.

A typical plant parasitic nematode examined under the high dry objective (430X) or oil immersion objective (900X) would reveal the following structures or organs (Fig. 6.1). At the anterior end is the *stoma* or *buccal cavity* with six lips bearing sensory papillae and a pair of pore-like or slit-like lateral sensory structures *(amphids)* near their base. Within the buccal cavity, one finds a darkened sharp protrusible *stylet* or spear with a tiny lumen for the passage of nutrients into the larger lumen of the esophagus. Stylets may differ in length, size, and shape according to the species, but typically they have three knobs at their base which serve as attachments for the protractor muscles. In some forms or males,

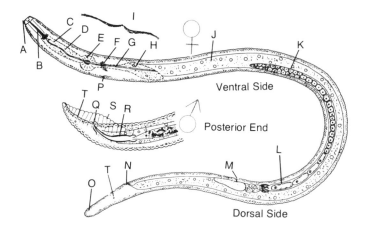

Ventral Side

Posterior End

Dorsal Side

(After Jenkins and Taylor)

FIG. 6.1. STRUCTURAL CHARACTERISTICS OF A PARASITIC NEMATODE
(PRATYLENCHUS)

A, Stoma; B, Stylet; C, Dorsal esophageal gland opening; D,
Procorpus; E, Metacorpus; F, Nerve ring; G, Isthmus; H, Basal
bulb; I, Esophagus; J, Intestine; K, Ovary; L, Maturing egg; M,
Vulva; N, Anus; O, Tail; P, Excretory pore; Q, Cloaca; R, Spicules,
S, Bursa; T, Phasmid.

the knobs may be reduced in size or entirely absent. Nonparasitic
nematodes generally do not possess stylets, but may have teeth or denti-
cles and show a wide range in the size and shape of the buccal cavity.

Below the stylet knobs, the lumen leads into and through the
esophagus, a muscular structure typically composed of three regions: 1)
the *corpus* which is subdivided into a narrow *procorpus,* and a bulb-like
metacorpus or *median bulb* with or without a valvular structure; 2) the
isthmus, a narrow region of varying length (partly encircled by a hardly
distinguishable *nerve ring* from which ganglia and nerves originate; and
3) a glandular *basal bulb* which terminates at the anterior end of the
intestine or may overlap it to a varying extent. In one group of nematodes
(the Tylenchoids), a *dorsal esophageal gland* empties into the lumen
just below the stylet knobs and may be located by an abrupt bend in the
lumen at this junction. The location of this junction as well as variations
in the shape and structures of the esophagus are important characteristics used in classifying nematodes.

The intestine of nematodes is essentially a hollow tube without con-
volutions extending posteriorly and ending with a slanted slit (the *anus)*
in the body wall near the tail. A supplementary excretory system, usually present in tylenchoid nematodes, may be observed as a canal system

emptying through a ventral pore in the cuticle. It is normally located ventrally in the region of the isthmus.

The tails of nematodes may be of various lengths and shapes. Their length is measured from the anal opening to the terminus. Nematodes lack visual organs and a definite respiratory or circulatory system. Respiration evidently takes place directly through the cuticle, and fluid circulation is accomplished by muscular movements within the body.

Nematode Reproduction and Growth

Both males and females are generally necessary for nematode reproduction but many exceptions that include hermaphroditism (male and female organs in the same individual) and parthenogenisis (reproduction without fertilization) have been noted. In some forms, males are absent altogether or, if present, have little or no role in reproduction.

The female reproductive organs consist of one (monodelphic) or two (didelphic) ovaries leading into an oviduct, a spermatheca and a uterus. When two ovaries are present, they may be parallel to each other or opposed (extending in opposite directions) and joined at their base into a common vagina and vulva which appears as a transverse slit-like opening in the cuticle on the ventral surface. The position of the vulva (V) varies with the species of nematode from near the midbody (V = 50–60%) in those with two ovaries, to near the end of the tail (V = 70–90%) in those with one ovary. In females that assume a spherical or pear shape, the vulva is found at the end of the body or close to it.

Male nematodes are characterized by a testis with spermatocytes leading to a spermatheca and vas deferens opening into a ventrally located cloaca, through which the forceps-like spicules (used in copulation) project. Another typically male structure present at the posterior end of some nematode species is a pair of fin-like cuticular extensions (bursa) also used in copulation.

Only mature males and females have fully developed reproductive structures. Those without them are juveniles (sometimes called larvae) and cannot be classified in keys to adults.

Reproduction of nematodes is usually oviparous (from eggs hatching after being laid) but may also be ovoviviparous (from eggs hatching from within the uterus). Eggs are generally oval and smooth. Plant parasitic nematodes may lay a few eggs per day to a maximum of several thousand in a two-month period. Eggs are deposited in the soil near or within roots, in flower heads, in infested leaf tissues or, in the case of cyst-forming nematodes, within the swollen body of the female whose dead body wall forms a protective cyst. From these cysts, second stage juveniles (larvae) emerge under the proper food stimulus or environmental conditioning, sometimes over a period of eight years. Fully grown embryos within eggs usually undergo one molt before emerging as second-stage infective juveniles.

Nematodes have no marked metamorphosis. Growth occurs through a series of molts, typically four in number. The first one or two may take place within the egg and the rest after hatching. In molting, the nematode sheds its cuticle completely and in addition, all cuticular-lined internal parts including the stylets. In juvenile Dorylaimoid nematodes, one or more new stylets can be frequently seen behind the terminal one.

Nematode growth is fairly rapid under warm summer temperatures, with the completion of a life cycle in about one month in most plant parasitic species (Fig. 6.2). Infective juveniles may live for long periods in the soil without food. Cool weather slows down growth and winter temperatures stop it altogether. Overwintering of nematodes in the cooler temperate zones, varying with the species, may take place in any stage in the soil or in plant tissues. Cyst-forming nematodes (Heterodera spp.) overwinter in the egg or juvenile stage within the cysts in the soil. Seed-gall nematodes (Anguina spp.) and the bulb and stem nematodes (Ditylenchus spp.) have the ability to undergo anabiosis as resting

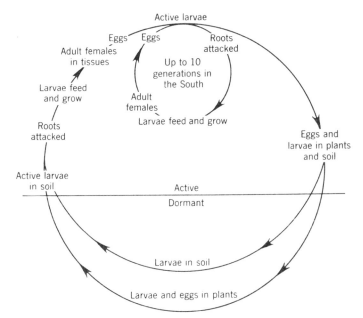

FIG. 6.2 THE LIFE HISTORY OF THE ROOT KNOT NEMATODE (MELOIDOGYNE)

wool-like masses of juveniles capable of surviving long periods, even years, of desiccation within seed galls or plant parts. Warmth and moisture miraculously bring them back to activity.

Dispersal of Nematodes

Plant parasitic nematodes move by a series of sinuous body undulations in water films around soil particles and plant parts. Their swimming movements are generally sluggish. On its own, a nematode moving in water films may travel no more than a foot or two in a single season. That is why initial infestations in a field produce very slowly expanding areas of plant damage. Mostly, however, they are disseminated over greater distances with running water or human assistance. They may be carried with soil, plants, or plant parts and transported from place to place when top soil, mulch, compost, potted plants, balled nursery plants or rooted cuttings are purchased; also with soil tillage equipment, rain water run-off, irrigation water, and field bagging materials. Plant parasitic nematodes are not known to pass through the intestinal tracts of humans, animals, or birds in a viable state, but may be carried from place to place on soil or mud attached to their feet.

Symptoms of Nematode Injury

Because most nematodes cannot be readily seen with the unaided eye and are difficult to isolate, they can build up to enormous populations in the soil before their presence produces noticeable symptoms in the plants attacked. Even when injury is noticed, the symptoms are seldom specific and can be easily confused with those caused by improper soil conditions, nutrient imbalances or deficiencies, lack of water, and disease caused by fungi, bacteria, and viruses. Some nematode species, even when not abundant enough to injure plants physically with their feeding, inject a toxic saliva which injures tissues. In addition, nematodes have been found to be indirectly responsible for the increase in severity of a number of root rots and wilts, since their feeding punctures allow ready access to fungi and bacteria. Dramatic reduction of the rots has been obtained when nematodes were eliminated from the soil. Nematodes have also been implicated as vectors of several soil-borne viruses.

The soil-inhabiting parasitic nematodes may be either migratory and *ectoparasitic* (contacting plant roots only when feeding) or *endoparasitic* (partly or wholly embedding themselves within plant tissues). The infective juvenile stages are always migratory.

The above-ground symptoms of plant injury are usually slow in showing up. Damage at first is not dramatic, and outright killing seldom occurs in one season. But as nematode populations build up, they slowly sap the plants' strength producing a decline in vigor and growth. Serious damage does not show up in temperate climates until warm soil temperatures have prevailed for some time and hot, dry periods have made it difficult for damaged or limited root systems to keep up with the water and nutrient requirements of the plants. Under such conditions, symptoms of attack may persist in spite of watering and feeding.

Above-ground symptoms of nematode infestation may appear as patches of poor growth in fields or plant beds, stunted or malformed new growth, yellowed or bronzed foliage on shrubs and trees, crinkled growth in a strawberry patch, elongated yellow lesions on narcissus foliage, swellings or knots on the stems and leaves of grasses, drought-susceptible areas on lawns, malformed and discolored seed heads in wheat fields, wedge-shaped, yellowed or brown areas on the lower foliage of chrysanthemums, foliage die-back and wilting during the hot part of the day.

Below-ground visual symptoms of nematode attack may appear as tiny discolored lesions on roots, shallow matted or reduced root systems, hairy root condition, malformed or stubby roots, decay in roots or bulbs, and knots, galls, or swellings on roots (Fig. 6.3).

Isolation of Nematodes

For a definite diagnosis, symptoms alone are not enough. Nematodes must be isolated from the soil or plant lesions and identified as parasitic before the symptoms may be attributed to them. Various methods are used to separate them from the soil or plant parts for identification under the microscope. Only a few simple methods are described.

Baermann Funnel Method.—This method is useful in isolating active parasitic and non-parasitic nematodes from the soil around plant roots. (Fig. 6.4).
Equipment
   One or more ring stands for 6 in.-glass funnels with a short large-bored stem.
   Two inch-long pieces of soft rubber tubing to fit over end of stems.
   Stiff spring clamps to pinch the ends of the rubber tubing and prevent water leakage.
   Rust proof hardware cloth (wire screening) 8–10 mesh, cut into 4½ in. discs to fit into funnel tops.
   Six-inch discs of muslin cloth or fine paper toweling to rest on wire screening and hold soil sample.

FIG. 6.3. ROOT KNOT NEMATODE INJURY TO TOMATO ROOTS

Procedure
1. Set up apparatus as indicated in Fig. 6.4.
2. Fill funnel with water to about ¼ in. of wire support.
3. Wet cloth or paper discs and place on wire supports.
4. Spread small handful of soil carefully and evenly on the cloth or paper.
5. Lift up one edge of cloth or paper carefully and add enough water to just flood the soil.
6. Wait 12–24 hr for nematodes to make their way through the screen and settle down to the pinchcock.
7. Place a small beaker under the rubber tube and open pinchcock to remove about 10–20 ml of water.
8. Distribute water in Syracuse watch glasses for examination for the presence of nematodes under a stereo-microscope.

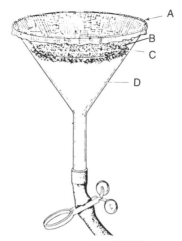

FIG. 6.4. BAERMANN FUNNEL
A, Unbleached muslin cloth or wet-strength tissue; B, Soil sample; C, Wire screen support; D, Water.

Isolation of Endoparasitic Nematodes.—The incubation-aeration method is probably the most efficient and effective method of isolating active endoparasitic nematodes from plant roots. Nematodes may be kept alive for weeks for future examination in aerated water.

Procedure

1. Place a handful of washed roots in a 500 ml beaker of water.
2. Cover the roots with water and aerate the water vigorously with a fish tank aerator or compressed air for two or more days. This tends to drive the nematodes out of the roots.
3. After elapsed period stop the aeration, remove the roots and allow about 10 min for the nematodes to settle to the bottom of the beaker.
4. Siphon off nematodes from the bottom with an eye-dropper or larger syringe and place sample in Syracuse watch glasses for examination in water under the stereo-microscope.

Soil Examination For Cyst-Forming Nematodes.—Fields may be surveyed for the presence of cysts of *Heterodera* spp. as follows:

1. Obtain ½ pt dry soil samples or allow soil samples to dry out.
2. Place a tablespoonful of the dry soil in a 500 ml beaker of water and stir.
3. Dry cysts (if present), weed seeds and debris will float to the top and collect around the periphery of the beaker. Skim off this material and place it in Syracuse watch glasses with a little water.
4. Examine carefully under the stereo-microscope for the presence of cysts.

Nematode Identification

Variations in both external and internal structures of soil inhabiting nematodes are used to classify them into eight groups which include both parasitic and nonparasitic forms. In general, the nonparasitic nematodes are much more active swimmers than the parasitic ones when observed in water through a microscope.

Rhabditoid Nematodes.—Nonparasitic; feed on soil microorganisms; stoma a sclerotized tube; esophagus with a muscular procorpus, a valveless metacorpus, long isthmus, and a basal bulb with a butterfly-like valve; 1 or 2 ovaries (V = 50–60% or V = 85–95%). (Fig. 6.5A).

Diplogasteroid Nematodes.—Nonparasitic; feed on soil microorganisms; stoma usually armed with a large dorsal tooth; esophagus with a valveless metacorpus, short isthmus, and a valveless basal bulb, V = 50%, commonly found in decaying vegetable matter and composted soil. (Fig. 6.5B).

Cephaloboid Nematodes.—Nonparasitic; feed on one-celled algae and bacteria; stoma a heavily sclerotized irregularly narrow opening; esophagus long with no metacorpus, with a rounded basal bulb containing a butterfly-like valve; V = 65%. (Fig. 6.5C).

Plectoid Nematodes.—Nonparasitic, very active; feed on one-celled algae and bacteria. Stoma elongate, sclerotized with an enlarged anterior portion; esophagus similar to cephaloboid group but valve in basal bulb elongated and somewhat spindle-shaped; V = 50%. (Fig. 6.5D).

Mononchoid Nematodes.—Nonparasitic, predatory; feed on other nematodes and small organisms; stoma large, rounded with a large forward-pointing tooth; esophagus cylindrical, muscular with no bulbs or valves; body up to several millimeters in length; V = 60%. (Fig. 6.6A).

Dorylaimoid Nematodes.—Parasitic and nonparasitic forms; nonparasitic forms possess a slanted-tipped stylet (odontostylet) with no basal knobs and a bottle-shaped esophagus with no valvular structures, (V = 40–55%). Juveniles may have one or more "spare" stylets below the functional one. Feed on algae, mosses and possibly plant roots but are not considered harmful. (Fig. 6.6B).

Plant parasitic forms belong to the genera *Xiphenema* (dagger nematodes), *Longidorus* (needle nematodes), and *Trichodorus* (stubby root nematodes). All three genera are migratory ectoparasites with bottle-shaped valveless esophagi. The first 2 genera contain nematodes up to 7 mm long with long needle-like stylets. The dagger nematodes have stylets with basal flanges and may produce root galls; juveniles may have a "spare" stylet; needle nematodes have no basal enlargements on their stylets; and stubby root nematodes, short and thick-bodied, have stylets curved like a bow posteriorly. (Figs. 6.8A, B, C).

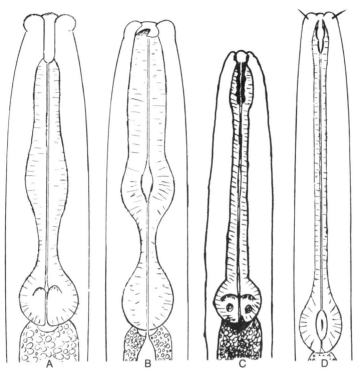

FIG. 6.5. ANTERIOR BODY STRUCTURE OF NEMATODE GROUPS
A, Rhabditoid; B, Diplogasteroid; C, Cephaloboid; D, Plectoid.

Aphelenchoid Nematodes.—Mostly parasitic—foliar nematodes, *Aphelenchoides* spp; and mostly nonparasitic—*Aphelenchus* spp. Parasitic forms generally endoparasites in leaves, flowers, and buds, very slender, up to 1.2 mm long, V =65–75%, active swimmers; stylets short, slender; *Aphelenchoides* spp. with very small basal knobs; knobs absent in *Aphelenchus* spp. Procorpus long and slender; metacorpus large and rectangular with a conspicuous valve; basal bulb glandular with long lobe overlapping anterior portion of intestine; dorsal esophageal gland outlet in metacorpus in both genera. The association of *Aphelenchoides* spp. and the bacterium, *Corynebacterium fascians* produces the "cauliflower disease" of strawberries. A closely related nematode, *Rhadinaphelenchus cocophilus*, produces the destructive "red ring" disease of coconuts in the Caribbean. (Fig. 6.6C).

Tylenchoid Nematodes.—The majority of destructive plant parasitic nematodes belong to this group. The general diagnostic structures are (Fig. 6.6D):

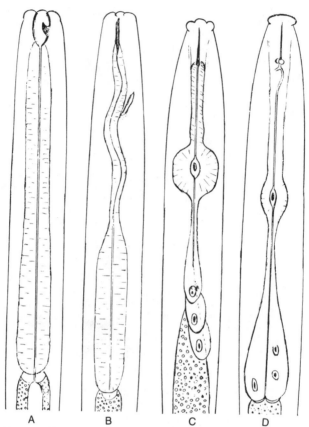

FIG. 6.6. ANTERIOR BODY STRUCTURE OF NEMATODE GROUPS
A, Mononchoid; B, Dorylaimoid (note developing stylet); C,
Aphelenchoid; D, Tylenchoid.

1. Well developed, heavily scleritized stylets with basal knobs.
2. An esophagus with a valvular metacorpus.
3. A distinct glandular valveless basal bulb which may or may not
   overlap the anterior portion of the intestine.
4. The outlet of the dorsal esophageal gland duct located in the lumen
   just below the stylet knobs on the dorsal side.

They may be endoparasitic or ectoparasitic and migratory or sedentary as mature females. Some 16 agriculturally important genera fall into
this group.

Cyst Nematodes (Heterodera *spp.*).—Distribution worldwide, endoparasitic; female body swells during growth and protrudes from root,
becomes distended with eggs; cuticle forms a tough cyst protecting the

eggs and first stage juveniles; cysts spherical, pyriform or lemon-shaped, white to brown in color; second stage juveniles the migratory infective stage; host range narrow to wide according to species.

Root Knot Nematodes (Meloidogyne *spp).*—Distribution worldwide, wide host range, endoparasitic; female body swells during growth,

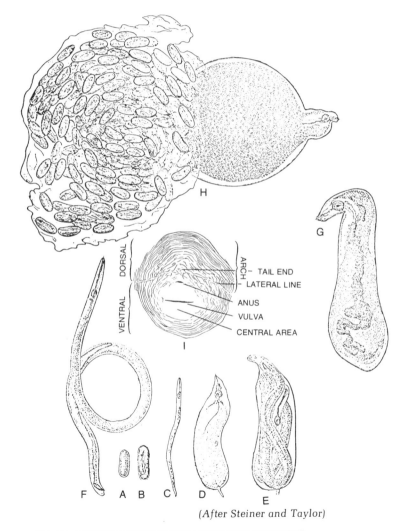

(After Steiner and Taylor)

FIG. 6.7. GROWTH STAGES OF THE ROOT-KNOT NEMATODE

A, Egg; B, First stage juvenile in egg; C, Infective juvenile; D, Third stage juvenile; E, Male within juvenile cuticles; F, Mature male; G, Young female; H, Mature female with egg mass; I, Perineal pattern.

becomes distended with eggs, eggs deposited externally in a gelatinous matrix; produce galls or swellings on the roots of plants; species separated on basis of host selection and differences in perineal pattern (cuticular finger-print-like pattern around the anus and vulva of the adult female) (Fig. 6.7).

Other Tylenchids in which the females become saccate (swollen) and sedentary and do not form a protective egg-cyst are grouped in the genera *Rotylenchulus* (reniform nematodes) found in a wide range of host plants in tropical and subtropical regions; *Nacobbus* (false root-knot nematode) found in western United States sugar beet fields; and *Tylenchulus* (citrus nematode) found throughout the citrus growing regions of the world and producing a slow decline of the plants.

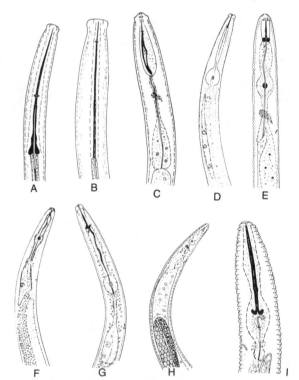

(*A, B, C, E, H, I After Jenkins and Taylor; D, F, G After Thorne*)

FIG. 6.8. ANTERIOR BODY MORPHOLOGY OF PARASITIC NEMATODE GENERA

A, Xiphenema; B, Longidorus; C, Tricodorus, D, Aphelenchoides; E, Hirschmanniella; F, Ditylenchus; G, Tylenchorhynchus; H, Anguina; I, Criconemoides.

Root-Lesion Nematodes (Pratylenchus spp.).—Distribution worldwide, very wide range of hosts; migratory endoparasites, small less than 1 mm in length, V = 70–80%, with short heavy stylets; produce stunted growth and small discolored lesions on roots, which facilitate entrance of root rot bacteria and fungi. (Fig. 6.1).

Burrowing Nematodes (Radopholus spp. and Hirschmanniella spp.).—Distribution mainly in tropical and subtropical regions; migratory endoparasites, long slender up to 2 mm in length, V = 50–60%. Radopholus spp. cause spreading decline of citrus, pepper yellows, "black head" of banana, and reduction of feeder roots in sugar cane. Hirschmanniella spp. attack mostly grasses and rice (Fig. 6.8E).

Bulb and Stem Nematodes (Ditylenchus spp.).—Distribution worldwide, migratory endoparasitic in stems, bulbs, tubers, leaves and flowers. Have ability to undergo anabiosis in pre-adult stage; length up to 1.4 mm, slender, V = 75–80%; races present, differing in host preferences, wide host range including fungi and algae; may cause bulb rots and root galls (Fig. 6.8F).

Stylet or Stunt Nematodes (Tylenchorhynchus spp.).—Distribution worldwide, migratory ectoparasitic; typical tylenchoid esophagus, non-overlapping basal bulb, V = 50–60%, length up to 1.4 mm. Wide range of hosts including cotton, peas, grasses, turf, tobacco, corn, sugarcane, rice, soy beans, sweet potatoes and herbaceous annuals, and woody perennials; produce stunted root systems. (Fig. 6.8G).

Seed and Leaf Gall Nematodes (Anguina spp.).—Distribution worldwide, mainly endoparasitic; length up to 4 mm, bodies rather stout, V = 85–90%. Induce gall formation in seeds, leaves, and other aerial parts of grasses and grains. Can undergo anabiosis in seed-like wheat galls as juveniles for up to 38 years of more. (Fig. 6.8H).

Ring Nematodes (Criconemoides spp., Criconema spp.).— Distribution worldwide, migratory ectoparasitic; very sluggish, short stout less than 1 mm long, with wide annules and long, strongly developed stylets; metacorpus with large ovoid valve, V = 85–95%. Not highly damaging to crops—woody perennials, turf, peanuts, tobacco, vetch, etc. (Fig. 6.8I).

Pin Nematodes (Paratylenchus spp.).—Distribution worldwide, mainly in temperate zones; migratory ectoparasitic, size minute slender up to .6 mm in length, V = 80–85%; wide host range (Fig. 6.9A). One very similar species, the Sessile Nematode (Cacopaurus), is found only on Persian walnut in California. Females very minute, up to ¼ mm long become obese, sedentary ectoparasitic.

Sheath Nematodes (Hemicycliophora *spp.*, Hemicriconemoides *spp.*).—Distribution worldwide, mainly in temperate and subtropical regions; migratory ectoparasitic, length up to 2 mm, V = 80–90%, adults with an extra cuticle (sheath). Hosts usually woody plants and grasses, also vegetable crops and other herbaceous annuals, small galls produced on root tips (Fig. 6.9C).

Lance Nematodes (Hoplolaimus *spp.*) *and Spiral Nematodes* (Helicotylenchus *spp.*, Rotylenchus *spp.*, Scutellonema *spp.*, Peltami-gratus *spp.*, *and* Aorolaimus *spp.*).—Distribution worldwide, in both warm and temperate climates; migratory ectoparasitic, semiendoparasitic, or endoparasitic; length .4–2.0 mm, bodies at rest spiral-shaped, c-shaped, or slightly curved; V = 50–60%, wide host range (Fig. 6.9B).

Awl Nematodes (Dolichodorus *spp.*) *and Sting Nematodes* (Belonolaimus *spp.*).—Distribution worldwide in tropic and temperate zones; migratory ectoparasitic; slim, up to 3 mm long; metacorpus with large valve, V = 50–60%; wide host range, in vegetable crops enter and attack germinating seed; in strawberries, peanuts, cotton, grapefruit, produce stubby root formation (Fig. 6.9D).

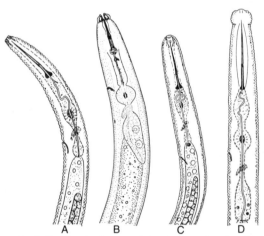

(A, C, D After Jenkins and Taylor, B After Goodey)

FIG. 6.9. ANTERIOR BODY MORPHOLOGY OF PARASITIC NEMATODE GENERA

A, Paratylenchus; B, Hoplolaimus; C, Hemicycliophora; D, Dolichodorus.

QUESTIONS FOR DISCUSSION
1. Why did many plant troubles caused by nematodes long remain unrecognized?
2. How may nematodes spread from one country to another? From one state to another? From one farm to another?
3. Under what conditions might a grower suspect nematode injury to his crop?
4. What makes nematodes hard to eradicate with chemical nematocides?
5. You suspect nematode injury to a crop. How should you proceed to try to verify it?
6. How do you differentiate between a plant parasitic nematode and a nonparasitic one? Between a mature female and immature one? Between a mature female and mature male?
7. How can some nematodes survive long periods of adverse conditions?
8. In what ways do plants react to the presence of parasitic nematodes?
9. Do swellings or galls on the roots of plants always indicate the presence of parasitic nematodes?
10. Some nematodes attack the foliage of plants. How do they get there?

Chapter 7

Plant Diseases

The study of plant diseases is known as plant pathology or phytopathology. It includes the study of the causes of plant diseases, their nature, and their control. When used in a broad sense, plant pathology includes plant disturbances attributable to all animals, large and small, and to all plants from the lowest to the highest forms, and all other factors that affect plant growth and crop production. In practice, however, the scope of plant pathology is generally limited to disturbances caused by microscopic organisms, viruses, and environmental factors (physiopaths). Insects, mites, rodents, and similar harmful organisms are covered in entomology or in economic zoology.

Plants, like animals, are beset by a multitude of diverse diseases caused by various living and nonliving factors. These diseases are expressed as differences from the normal appearance or productiveness of the plants. A diseased plant may be defined as one that is continuously being irritated by a causal factor that upsets the normal activity of cells or organs so as to produce visible evidence of disease in the form of characteristic symptoms. This differs from an injury, which is only a momentary irritation resulting in injurious effects.

Plant pathology is a relatively young science. Plant diseases themselves, however, are not of recent origin. References to blights, mildews, and plagues have been found in the earliest records of man. Throughout human history great plant disease epidemics (enphytotics) that brought famine and disaster to millions and that influenced the migrations of peoples have been recorded. The great famine of 1845 in Ireland and European countries was caused by an enphytotic of the late blight fungus on potatoes. Because plant pathology is a relatively new science, there is still some confusion in the standardization of terms.

No plant species or variety exists that is free from attack by disease. Most plants are susceptible to many disease-producing organisms, some of which may, under favorable conditions, wipe out the plants in local areas or even throughout their range. Our native chestnut trees are now only a memory of our fathers in most areas because of the destructive chestnut blight first observed in the New York Zoological Park in 1904. Some specific disease-producing organisms attack a wide range of different plant species, others attack only closely related species of plants, and still others are limited in the attacks to one or more definite varieties of the same species.

Healthy Versus Diseased Plants

When is a plant diseased and when is it healthy? Generally a plant is considered healthy if it has the desired inherent traits; if it is grown under the best conditions with respect to soil, nutrients, moisture, air, and light; and if it is free from any insect attack, disease attack, and injury. Any departure from the optimum of any one or more of the above factors tends to produce a sick plant. Since plant species vary greatly in some of their requirements, the need for being well informed on the requirements of different crops is evident.

Various degrees of being sick are evidenced by plants. This depends how far from the ideal any one or more of the factors for health is removed. Since most plants do not grow under ideal conditions, they might all look diseased to a varying degree if compared with specimens grown under absolutely optimum conditions for growth or crop production. Since such an ideal comparison is impracticable, the more obvious symptoms of disease, such as abnormality of growth, color, habit, or crop, must be sought. The importance of knowing a plant's normal habits or requirements in order to recognize any abnormalities cannot be overemphasized.

The Nature of Disease Attack

A plant disease is generally characterized by a definite series of symptoms or symptom complexes that are expressions of that disease. Like any of our own common ailments, a plant disease does not generally produce one single symptom but produces a progressive series of symptoms, sometimes on different organs of the same plant. As an example, the brown rot of stone fruits may first appear on peaches as a blossom blight. From the blossoms the disease may travel into the twigs and produce girdling cankers that wilt and later kill the growth above them. Later in the summer, the fruit may be attacked just as it is ripening to produce a rot that invades the entire fruit. The fruit may remain clinging to the tree, may gradually shrivel, and may become a mummy, from which the cycle may start all over again the next spring. A single disease attacking different plant species or different organs of the same plant may produce a variety of symptoms all forming a symptom complex characteristic of that disease.

Plant diseases may be either localized or systemic. A localized disease is one in which the causal organism is confined to definite organs or areas and does not spread internally to other organs of the plant. Leaf spots, scab, and fruit rot diseases are good examples of localized diseases. In a systemic disease the causal factor tends to spread internally and affect many parts of the plant. Most virus diseases are systemic.

Some bacterial diseases such as fire blight and some fungous diseases such as fusarium wilt tend to become systemic.

The Diagnosis of Plant Diseases

The health of a crop is dependent upon the grower's skill and knowledge. The grower must recognize at once when something is wrong and make a diagnosis before he can apply control measures intelligently. Diagnosis consists of telling diseases apart from a knowledge of their characteristic *signs* and *symptoms*. It is an art that requires constant experience and cultivation. Growers should have a knowledge of the general symptoms of the common diseases of their crops. Symptoms in many cases are so complex and confusing that the services of a trained plant pathologist are necessary to determine their cause correctly.

To prove that a disease is caused by a certain infectious organism, the plant pathologist may apply *Koch's Postulates* defining the evidence:
1. The organism should be associated with all cases of a given disease.
2. The organism must be isolated from the diseased plant.
3. When the organism is subsequently inoculated into susceptible plants, it must reproduce the disease.
4. The organism must be re-isolated from the experimental infection.

All these postulates cannot be fulfilled with every disease but in most cases there may be sufficient evidence to link symptoms with a specific pathogen.

Symptoms are evidence of disease expressed by the plant itself because of the irritation or presence of some causal factor. Physicians generally make their diagnoses through symptoms—it is quicker and easier. Growers or plant doctors may also make their diagnosis through symptoms, but when diseases are difficult to diagnose by this means, plant doctors, like medical docotors, search for *signs*—the causal organisms or factors producing the symptoms. Symptom complexes characteristic of a disease, together with any accompanying signs, constitute the effect and cause of a disease on which accurate identification is based. Such common symptoms as spots in leaves; dead areas in bark; rotten areas in fruits or tubers; abnormal swellings on roots, branches, and foliage; and abnormal color or habit are all expressions of the reaction of the plant attacked to the irritation of the disease-inducing factor.

The Classification of Plant Disease Symptoms

Symptoms may be either *lesional* or *habital*. Lesional symptoms are those confined to definite areas, and they show structural changes known as lesions in the epidermis or bark. Leaf spots, galls, cankers, and rots are examples of lesional symptoms. Habital symptoms are unnatural or abnormal habits that result from generalized disease effects. Abnormal foliage color, wilting, curling, and stunting are examples of habital symptoms.

Symptoms may be either *primary* or *secondary*. Primary symptoms are those occurring on the plant at the point of attack of the organism. Secondary symptoms are those occurring on some other part or parts of the plant as a result of the irritation or injury produced at the primary point of attack. For example, in crown gall the large, knotty growths on the roots or crowns of raspberries and other fruits are primary symptoms, but the stunting of the plants and the yellowed foliage resulting from these growths are secondary symptoms. The canker caused by brown rot on a peach twig is a primary symptom, whereas the wilting of the foliage and the death of the twig above this canker is a secondary symptom. It is very important in disease diagnosis to differentiate between primary and secondary symptoms. Primary symptoms should always be sought if symptoms are suspected of being secondary. Yellowed leaves, for example, may be either a primary or a secondary symptom of disease. If the causal factor cannot be located on or in the leaves after careful examination, it is reasonable to suspect that this is a secondary symptom. The primary symptom or factor involved should be sought elsewhere on the plant or in its environment.

Types of Symptoms

Symptoms of plant disease fall into four main types: (1) those in which the outstanding characteristic is the dying or death of cells, tissues, or organs (*necrotic symptoms*); (2) those that are characterized by the loss, suppression, or reduction of the normal green color in plants (*chlorotic symptoms*); (3) those in which the predominating characteristic is underdevelopment of or stopping of growth of cells, tissues, or organs (*hypoplastic symptoms*); and (4) those that feature overdevelopment of cells, tissues, or organs (*hyperplastic symptoms*).

During its course, a disease may exhibit symptoms that fall into any one or more of the four classifications, since all disease processes eventually tend to end with death of the whole plant or some of its parts. Care must be taken in classifying a disease to select the predominating reaction as chlorotic, necrotic, hypoplastic, or hyperplastic. It is often difficult to decide which of the responses is the dominant one, and differences of opinion may be expected.

Necrotic, chlorotic and hypoplastic symptoms may be localized in certain organs or parts of them, or they may be general, resulting in the death or stunting of the whole plant. Hyperplastic symptoms are usually primary symptoms of a lesional character, but in some cases they may be habital. Hyperplastic types of diseases may, as they progress, show secondary hypoplastic, chlorotic, or necrotic symptoms.

Some Common Necrotic Symptoms (Necroses).—*Hydrosis.*—A water-soaked condition of tissues caused by water leaving the cells and filling up the intercellular spaces. This is the first symptom to be noted in many foliage diseases such as late blight of potatoes and tomatoes (Fig. 7.1).

Wilting.—The flaccid condition of leaves or shoots caused by a loss in turgidity (Fig. 7.2). This is usually a secondary symptom due to some trouble in the conducting system of the plant or in the soil. Wilting may be permanent, resulting in death, or temporary, with the plants recovering at night.

Scorch (firing).—The sudden death and browning of foliage or sometimes fruit (Fig. 7.3). Scorches are caused by physical factors such as drought, excessively high temperatures, and injurious chemicals.

FIG. 7.1. HYDROSIS CAUSED BY THE LATE BLIGHT FUNGUS OF POTATO
Note white fungous growth around lesion.

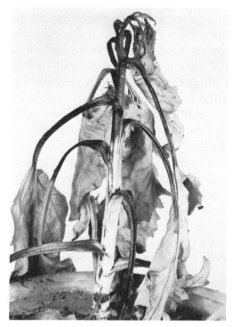

FIG. 7.2. WILTING OF ASTER PLANT CAUSED BY A *FUSARIUM* FUNGUS
Note white mold and blackened stem typical of advanced cases.

FIG. 7.3. LEAF SCORCH, CAUSED BY DROUGHT AND HIGH TEMPERA-
TURES, ON DOGWOOD

Blight.—Similar to scorch but is usually applied to foliage or blossoms killed by a disease organism (Fig. 7.4).

Die-Back.—The dying backward from the tip of foliage, twigs, or branches of trees and shrubs.

Spot.—A term applied to more or less discolored circular or angular lesions on foliage, fruit, or stems (Fig. 7.5). Spots are sometimes characterized by target-like banding or zonations of various colors. When the diseased tissues in spots drop out of leaves, leaving circular holes, the term *shot-hole* is applied.

Rot.—Dead tissue in a more or less advanced stage of decay (Fig. 7.6). It is generally lesional and may appear in woody organs as well as in fleshy ones on all parts of a plant. Thus, there may be wood rots, stem rots, bud rots, fruit rots, and root rots. In many fruit rot diseases the final phase is *mummification*, the shriveled up, dry, fruit mummies that may remain clinging to the branches.

Canker.—Localized, sharply demarked dead lesions, usually sunken

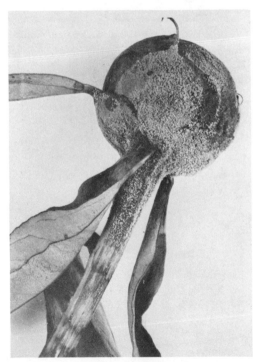

FIG. 7.4. BUD BLIGHT CAUSED BY *BOTRYTIS* BLIGHT FUNGUS OF PEONIES

Note typical mold on affected area.

FIG. 7.5. ANTHRACNOSE OF RASPBERRIES PRODUCES SPOTS ON STEMS

FIG. 7.6. BROWN ROT OF CHERRY
Note mummy being formed.

or cracked, in the bark of the trunk, stems, and roots of trees and shrubs or, rarely, on herbaceous plants (Fig. 7.7). Perennial cankers may show concentric ridges of callus growth.

Damping-Off.—The rapid rotting of seeds or seedlings before they emerge from the ground (pre-emergence damping-off) or the rapid rotting of the bases of seedlings so that they fall over (post-emergence damping-off).

Bleeding.—The continuous oozing of sap from wounds or injuries. When this bleeding is accompanied by fermentation of the sap, it is referred to as *slime-flux*; when it is gummy as on citrus or stone fruits, it is called *gummosis*; and when it occurs in coniferous trees, it is called *resinosis.*

Scald.—The blanching of epidermal tissues that gradually turn pale brown or darken. This symptom commonly occurs on fruits, sometimes on foliage (Fig. 7.8).

Some Chlorotic Symptoms.—Some textbooks eliminate this category and place chlorosis under hypoplastic symptoms and yellowing under necrotic symptoms (Roberts and Boothroyd 1972).

Chlorosis.—The partial failure of the development of green color in organs that are normally green. Chlorosis may be diffused or patterned.

FIG. 7.7. CANKER CAUSED BY *NECTRIA* SPP. ON ALBIZZIA

Note pruning wound through which infection started.

FIG. 7.8. SCALD ON TOMATO CAUSED BY HEAT FROM SUN

In the diffused type, the foliage is light green or yellowish green and is subnormal in size. Patterned chlorosis (*mosaics*) is characterized by the mottling of leaves or fruit with definite areas of normal green, light green, and yellow (Fig. 7.9). Inheritable chlorosis is portrayed by some varieties of arborvitae and a variety of yew in which all or part of the foliage is yellowish.

Yellowing.—The yellowing of normally green tissues (usually foliage) is due to a breakdown of chloroplasts or chlorophyll. When yellowing is generalized, involving all the foliage of a plant, it is usually a secondary symptom and may be caused by a disease-producing organism that has attacked some other important organs or by an unfavorable environmental condition.

Some Common Hypoplastic Symptoms (Hypoplases).—*Albication.*—The complete suppression of green color in normally green tissues.

Suppression.—The failure of certain organs such as flowers or seeds to develop at points where they normally develop.

Etiolation.—Excessive spindliness accompanied by dwarfing of foliage and flowers and lack of natural chlorophyll development.

Dwarfing or Stunting.—A habital symptom characterized by the subnormal size of the whole plant or of some of its organs. The symptom may be primary if growth halts in the area of attack or secondary if the whole plant is or if some of its organs are subnormal in growth.

Rosetting.—The crowding of the foliage into a rosette formation caused by the shortening of the internodes of stems or of branches.

FIG. 7.9. TOMATO MOSAIC; LEAVES ON LEFT NORMAL

Some Hyperplastic Symptoms (Hyperplases).—*Proplepsis.*—The development of shoots from buds that normally remain dormant. This symptom commonly follows die-back.

Abscission or Defoliation.—The premature falling of leaves, flowers, or fruits caused by the formation of the abscission layer at the bases of their petioles. The abscission layer forms normally in the fall of the year, causing fruit and foliage drop from deciduous plants. On evergreens, the abscission layer forms on old leaves and needles in the late spring or summer, causing them to drop. Defoliation is a secondary symptom of many diseases.

Russeting.—The brownish, superficial roughening of the epidermis of fruits, tubers, and other structures. Injury to the growing epidermis causes suberization and the consequent roughening of the affected areas.

Discoloration.—The abnormal coloration of tissues or organs other than yellowing or chlorosis. Discoloration may be lesional or diffused and is frequently a secondary symptom of disease attack. A coppery tinge in the epidermis of green leaves is referred to as *bronzing*, and a purplish or reddish tinge in green leaves or other plant parts caused by the development of anthocyanin pigments is termed *purpling*. Purpling may be localized around lesions or may be diffused in the organs as a secondary symptom of injury to roots or branches.

Galls (Tumefaction).—Local swellings or tumor-like growths on any part of a plant (Fig. 7.10). They may be large or small, smooth or rough, and fleshy or woody. They may bear descriptive names such as warts, knots, tubercles, boils, blisters, and clubs.

Callus.—New tissue growing over a wound in an attempt to heal it. Callus formation is normal around a pruning wound (Fig. 7.11), but it often appears around cankered areas as a typical symptom.

Curling (Rolling).—The abnormal curling or bending of shoots or leaves because of tissue overgrowth on one side of the affected organ (Fig. 7.12). Rolling or curling downwards is the first symptom of thirst in leaves.

Scab.—Definite, slightly raised and roughened lesions (sometimes cracked or corky) on fruits, tubers, leaves, or stems (Fig. 7.13). Scab diseases are lesional.

Fasciculation.—The abnormal development of twigs or roots about a common point. When this occurs on the aerial parts of a plant it is usually called a witches' broom (Fig. 7.14).

FIG. 7.10. BLACK KNOT ON A PLUM CAUSED BY A FUNGUS

FIG. 7.11. A WOUND WITH CALLUS GROWTH AROUND IT
Note also a completely healed wound directly above.

FIG. 7.12. PEACH LEAF CURL CAUSED BY A FUNGUS

FIG. 7.13. APPLE SCAB CAUSED BY A FUNGUS ON YOUNG FRUIT AND ON LEAF

FIG. 7.14. WITCHES BROOM ON BLUEBERRY CAUSED BY A RUST FUNGUS

Disease Products

These are exudates and odors that originate from both the disease-producing organism and from its host or from the disease-producing organism alone. *Ooze* is the term given to sticky liquid exudates from lesions containing gelatinous-walled bacteria or spores of fungi mixed with the disorganized products of the plant. Ooze usually forms as drops on the surface of lesions as in the case of a fire blight canker, as the ooze from the cut stem of a cucumber plant affected with bacterial wilt, or as a fermenting exudate from tree trunks called *slimeflux*. Many diseases produce a characteristic odor that indicates their presence in the plant or plant part attacked. For example, the late blight of potatoes, the caratovorus soft rot of vegetables, and the orange rust of cane fruits give off characteristic odors.

QUESTIONS FOR DISCUSSION

1. Why is it harder to determine when a plant is diseased than when a person is diseased?
2. Why can most plants be considered diseased to some degree?
3. Is it possible for the layman to mistake a healthy plant for a diseased one? Vice versa?
4. Why must one know what a normal plant of the species looks like during its growth before one can recognize a diseased condition?
5. In what ways does a plant disease resemble a human disease?
6. Are most plant diseases systemic or localized? Human diseases?
7. Do you think growers should be able to diagnose the common diseases of their crops, or should they rely on a plant pathologist?
8. Which is a more reliable basis for disease diagnosis: signs or symptoms?
9. Why is it important to recognize whether a symptom is primary or secondary?
10. Yellow foliage on normally green plants may be either a necrotic symptom or chlorotic symptom. What determines the classification?
11. Die-back most often occurs in the late winter and early spring. Why?
12. Can rots occur in plant organs without the presence of bacteria or fungi?
13. Sun-scald cankers usually occur on the southwest side of trees. Why?

14. Why is defoliation considered a hyperplastic symptom?
15. Why is the curling of shoots considered hyperplastic?
16. Is it necessary to diagnose a disease correctly in order to be able to control it?
17. Are plants able to overcome some of their diseases without human help?

Chapter 8

Causes of Plant Diseases

The study of the cause of diseases is generally referred to as etiology (the science of cause). In plant pathology, it deals with the primary causal agents or signs of disease in plants and their relation to the suscept or host.

These primary causal agents, generally referred to as *pathogenes*, fall into three distinct groups: (1) living organisms, (2) viruses, and (3) physiopaths—nonliving factors such as improper nutritional and environmental conditions. Viruses and physiopaths are discussed in Chapter 10.

LIVING ORGANISMS CAUSING DISEASE

Diseases caused by living organisms and viruses are usually termed infectious diseases. The attacking organisms are the parasites, and the plants attacked are usually called hosts or suscepts. The living organisms responsible for plant diseases are found mainly in the groups known as bacteria (Schizomycetes), fungi (Eumycetes), and nematode worms (Nemata) which are discussed in Chapter 6.

In many instances, disease-producing organisms spend part of their life cycle in living tissue and part in dead tissue. Plant pathogenes such as the powdery mildews and the rusts that depend entirely on living plant tissues for food are known as *obligate parasites*. Some plant pathogenes have become so specialized that they can attack only certain varieties of plant species and are classified as physiological strains or races.

BACTERIA (Schizomycetes)

Bacteria are the cause of many destructive plant diseases such as corn wilt, cucumber wilt, fire blight of pears, soft rot of vegetables, crown gall of fruits, and black spot of delphinium. They are minute, one-celled organisms without chlorophyll but are nevertheless regarded as primitive plants. Although many species were known to be pathogenic to man and to animals, they were not known to cause plant diseases until Burrill discovered about 1879 that fire blight of pears was caused by a member of this group. Since then, nearly 200 bacterial organisms that cause diseases of flowering plants have been found.

Classification of Bacteria

Bacteria are grouped into three types—the coccus forms, the bacillus forms, and the spirillum forms. Practically all of those that cause plant diseases fall into the bacillus or rod-shaped group (Fig. 8.1) order Eubacteriales, genera *Erwinia, Pseudomonas, Xanthomonas, Agrobacterium* and *Cornebacterium*. A few plant pathogenic bacteria belong to the higher order Actinomycetales, genus *Streptomyces*. The most important is *Streptomyces scabies*, the cause of common potato scab. This genus is characterized by an extremely fine filamentous branched body with spiral branches bearing spores (conidia) in chains, a characteristic of many fungi. The plant pathogenic bacilli are differentiated by the number and location of flagellae, their physiological characteristics and appearance when grown in different agar cultures, and their reaction to the Gram stain (gram-negative or gram-positive). The system of naming bacteria has undergone several changes in the past, and as a result the same organism may be known by two or more scientific names.

Only a few of the plant-pathogenic bacteria are known to form spores; they multiply rapidly in host tissue by simple division. Most are gram-

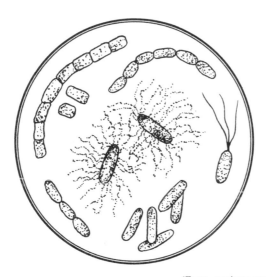

(From various sources)

FIG. 8.1. BACILLUS TYPES OF BACTERIA

negative; that is, they do not take on the standard Gram stain. They are generally white or yellow, motile with flagella or nonmotile and aerobic. Some are obligate parasites; that is, they depend on continuous association with living tissues of their hosts for existence; others may live for various lengths of time as saprophytes on dead plant tissues, in the soil, or in culture media. They are favored by high temperatures.

Mode of Attack

Succulent plant tissues are most susceptible to bacterial attack. Some bacteria can enter only through wounds or lesions made by natural forces or by insects or other animals; others may gain entrance through water pores, stomata, lenticels, and in a few cases the nectaries of flowers. Bacteria cannot enter directly through the epidermis. Once inside the host's tissue they multiply and migrate from cell to cell and may invade the vascular system to become systemic.

Dissemination

Bacterial disease organisms may survive from year to year in seeds, plant parts, soil, or the intestines of insects. They may be spread or disseminated by insects, splashing rain drops, dripping dew, running water, animals, tools, plant parts, or infected seed. Symptoms of a bacterial disease may not differ greatly from those of other diseases, and therefore accurate diagnosis may require that the causal organism be found in the lesion or in the ooze that may form on or in the affected tissues.

Pathological Effects

Bacteria may harm plants by secreting enzymes that destroy cells, by secreting toxic substances that poison and kill host cells, or by producing hormones that lead to cancerous cell development. Bacterial diseases may produce three types of disease symptoms in plants: (1) localized or generalized killing of tissues, as found in blights, spots, and rots; (2) invasion and spread through the vascular tissues, producing wilts; and (3) excessive multiplication of host cells, producing galls.

THE FUNGI

Fungi are primitive plants, but they lack chlorophyll and must therefore obtain their food from animals or plants. There are thousands of species of fungi with seemingly endless adaptations and diversifications. Most are harmless saprophytes competing with and destroying harmful organisms, feeding on dead animal or plant remains, and help-

TABLE 8.1

SOME IMPORTANT BACTERIAL PLANT DISEASES

Disease	Cause	Host
Alfalfa wilt	Cornebacterium insidiosum	Alfalfa
Angular leaf spot of cotton	Xanthomonas malvacearum	Cotton
Bacterial wilt of corn	Xanthomonas stewartii	Corn
Bacterial wilt of cucurbits	Erwinia tracheiphila	Cucurbits
Bacterial blight of beans	Xanthomonas phaseoli	Beans
Black rot of crucifers	Xanthomonas campestris	Crucifers
Crown gall	Agrobacterium tumefaciens	Fruits, cotton, alfalfa, etc.
Fire blight	Erwinia amylovora	Pome fruits
Potato scab	Streptomyces scabies	Potato
Ring rot	Cornebacterium sepedonicum	Potato
Soft rot	Erwinia carotovora	Root vegetables, cabbage, lettuce
Wilt	Pseudomonas malvacearum	Cotton

ing to break down organic matter. Some are used for making cheese and for producing mushrooms for human consumption; others parasitize insects. A few cause diseases of man, such as "athletes' foot" and "ringworm." Many fungi invade the tissues of higher plants and these produce the majority of infectious plant diseases. During their life, fungi (like other plants) first produce vegetative structures, which may later give rise to reproductive structures.

Vegetative Structures of Fungi

The body of a fungus (the *thallus*) typically consists of delicate tubular, cottony filaments known as *hyphae* (sing. *hypha*), which form branched systems of threads called *mycelia* (Fig. 8.2); however, some fungi are one-celled or amoeboid. When the mycelium forms as a dense, or loose, woolly mat or web, it is usually referred to as a *felt*. Hyphae are generally white but may be red, black, brown, yellow, olive, or greenish. Aerial hyphae, such as those found in the powdery mildews, send specialized absorbing organs, *haustoria*, into the cells. Intercellular hyphae such as those found in the rusts may also send haustoria into the cells to obtain nourishment.

The mycelia of fungi may exhibit different branching habits and structure, which help in identification and diagnosis. When a mycelium begins to grow, it generally radiates from its point of origin and therefore

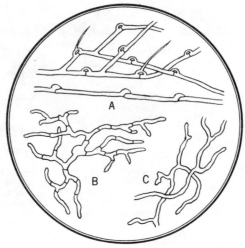

(From various sources)

FIG. 8.2. THE MYCELIA OF FUNGI

A, Basidiomycetes type of mycelium showing clamp connections; B, Ascomycetes type of mycelium showing its septate character; C, Phycomycetes type of mycelium showing its non-septate character.

tends to produce round lesions, characteristic of many fungous diseases.

Some fungi excrete toxic substances into the surrounding tissues, destroying them, and then live on the dead tissues as saprophytes; others feed directly on living cells, absorbing the cellular food through the hyphal walls or through the haustoria.

The Reproductive Structures of Fungi

Fungi commonly disseminate and reproduce themselves by means of tiny, microscopic bodies called *spores* (Fig. 8.3), which function like the seeds of higher plants. A spore landing on the appropriate part of a suscept under proper moisture and temperature conditions is capable of producing a new fungous body. These spores are named in accordance with the way in which they originate. If they are produced either by budding or by breaking off from the tips of specialized hyphae, they are called *conidia* and the hyphae are called *conidiophores* (Fig. 8.4). If the spores are produced within or on the outside of structures known as spore fruits or fruiting bodies, they are given names depending on the type of fruiting body from which they originate. Spores of different fungous species vary greatly in size, shape, color, and origin. Often their presence on or in a lesion can be used as a diagnostic sign of disease.

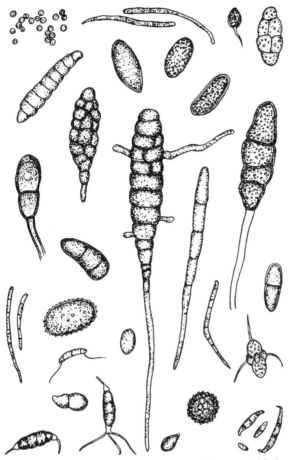

(Redrawn from Schwarze, N.J. Agr.
Expt. Sta. Bull. 313)

FIG. 8.3. THE SPORES OF FUNGI SHOWING THE GREAT VARIATIONS IN
THEIR STRUCTURE, ALL GREATLY MAGNIFIED

Spores vary in size from 1μ to 1 mm, but generally, unless seen in aggregate masses, they are not distinguishable by the naked eye.

Types of Spores.—Spores may be *motile* or *nonmotile*, and *sexual* or *asexual*. The same fungus may produce several types of spores during its seasonal cycle. Most fungi produce 2 or 3 types of spores, but some, like the rusts, may have as many as 5 types. Spores are usually produced on the surface of infected tissues, but sometimes spores form within plant tissues. Spores may be one-celled, two-celled or several-celled. They are produced sexually as the result of the fusion of two similar or dissimilar

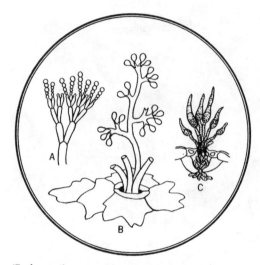

(Redrawn from Heald, Manual of Plant Diseases)

FIG. 8.4. CONIDIOPHORES

A, Blue mold or *Penicillium;* B, Downy mildew (Peronospora); C, Cercospora.

cells (Figs. 8.5 and 9.9) or asexually. When spores are motile they are usually called swarm spores or zoöspores (Fig. 8.6). In all downy mildews except species of Peronospora, swarm spores are produced. They swim around for a brief period, encyst, then germinate by the production of a germ tube.

Special resting or resistant spores that favor survival of the fungus are called *chlamydospores* when formed simply as enlarged, thick-walled hyphal cells. They may occur singly or in groups in the hyphae.

Spores may be disseminated by air currents, splashing rain, running water, animals, tools, and man.

The fruiting bodies that bear the spores are discussed under the different groups of fungi of which they are characteristic.

Sexual reproduction in fungi typically consists of three distinct phases:

1. *Plasmogamy*—The union of two sex cells which brings one haploid nucleus from each cell close together in one cell.
2. *Karyogamy*—The fusion of the two nuclei to produce one diploid zygote nucleus.
3. *Meiosis*—The reduction division of the diploid nucleus which restores the haploid condition in the nuclei of the spores and the new fungus body.

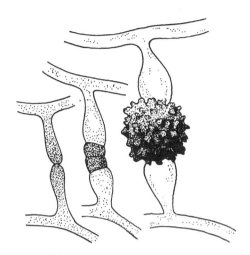

FIG. 8.5. STAGES IN THE DEVELOPMENT OF A SEXUAL SPORE (ZYGO-SPORE) AS THE RESULT OF THE UNION OF TWO SIMILAR CELLS (GAMETANGIA)

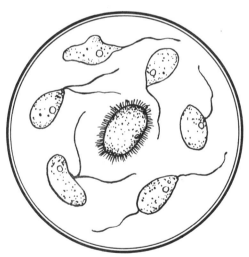

FIG. 8.6. TYPES OF SWARMSPORES OR ZOÖSPORES

In many of the lower fungi, karyogamy follows plasmogamy almost immediately; however, in the higher fungi the two sexual nuclei brought together in one cell may remain separated as a *dikaryon* for a considerable length of time with daughter cells retaining the dikaryotic condition until karyogamy takes place.

The asexual or imperfect stage (the *gametophyte*) of a fungus is the haploid (n) stage, whereas the sexual or perfect stage (the *sporophyte*) is the diploid (2n) stage. Asexual reproduction is usually repeated several times during the growing season and may result in severe disease damage to crops under conditions favorable to the pathogen; whereas the sexual stage of many fungi is produced but once a year, generally in the spring, and produces the initial infection of a crop.

The Classification of Fungi

On the basis of structures and methods of reproduction, fungi may be divided into five agriculturally important classes: (1) *Plasmodiophoromycetes*, the primitive fungi, or endoparasitic slime molds; (2) *Phycomycetes* (sub-class *Zygomycetes*—the algal fungi and sub-class *Oomycetes*—the egg fungi); (3) *Ascomycetes*, the sac fungi; (4) *Basidiomycetes*, the club fungi (sub-class *Heterobasidiomycetes*—the rusts and smuts and the sub-class *Homobasidomycetes*—the mushrooms and shelf fungi); and (5) *Deuteromycetes* (Fungi Imperfecti), the imperfect fungi and the form-order *Mycelia Sterilia* with no reproductive spores.

The Plasmodiophoromycetes.—This is a primitive class of fungi that resembles the nonparasitic slime molds (Myxomycetes) in its naked amoeboid thalli (*plasmodia*) and the Phycomycetes in its manner of spore formation. Thick-walled resting spores, *hypnospores*, are formed by the fragmentation of whole thalli in a process called *holocarpic reproduction*. Hypnospores germinate to produce zoöspores which fuse like gametes and invade the roots of the suscept to produce new plasmodia. The root invasion causes the abnormal growth and enlargement of the host cells (tumefaction). There are only two species of economic importance: *Plasmodiophora brassicae*, the cause of clubroot of crucifers, and *Spongospora subterranea*, the cause of powdery scab of potatoes.

The Phycomycetes.—The Phycomycetes include many forms that are pathogenic to higher plants. Generally, the mycelium is multinucleate, single-branched, and lacking in cross walls (nonseptate) except those that set off the reproductive organs (Fig. 8.2C). Septations sometimes occur in young hyphae or in old hyphae in the process of forming chlamydospores.

Asexual reproduction in most Phycomycetes takes place through the production of sporangia (Fig. 8.7) that bear either simple, air-borne spores or motile, water-borne swarm spores (zoöspores). Conidia may be produced in some cases.

Sexual reproduction, when present, is either by the union of the two unequal sex cells or gametes to form an *oöspore* (Fig. 9.9) as in the Oomycetes, or by the union of two equal sex cells or gametes to form a *zygospore* (Fig. 8.5) as in the Zygomycetes.

TABLE 8.2

SOME OF THE MOST IMPORTANT DISEASES CAUSED BY PHYCOMYCETES

Common Name of Disease	Causal Organism	Host Plants
Damping-off and root rots	*Pythium debaryanum* and other species	Numerous species
Late blight	*Phytophthora infestans*	Potato, tomato
White rusts	*Albugo candida* and other species	Crucifers, *Compositae*, sweet potato, salsify
Brown spot of maize	*Physoderma zea-maydis*	Corn
Red stele	*Phytophthora fragariae*	Strawberries
Black wart of potato	*Synchytrium endobioticum*	Potato
Downy mildews	*Sclerospora, Plasmopara, Pseudoperonospora,* and *Bremia*	Grains, grasses, onion, grapes cucurbits, crucifers,
Soft rot, leak, vegetable rot	*Rhizopus stolonifer* and other species	Sweet potato, various vegetables and fruits in storage

The Ascomycetes (Sac Fungi).—The higher Ascomycetes all have the common characteristics of a septate mycelium and of forming sexual spores (*ascospores*) in a specialized sac-like cell or *ascus*. They have no motile spore forms.

Vegetative Structures.—The hyphae are generally septate and composed of uninucleate cells (Fig. 8.2B). Haustoria are present in those forms known as powdery mildews and sooty molds. Some hyphal cells may separate with age and form conidia, or they may become thick-walled and form *chlamydospores*. In some Ascomycetes the mycelium may aggregate into compact masses called *sclerotia* or *stromata* (Figs. 8.8 and 8.9). In this stage the fungus is able to survive long periods of unfavorable conditions. In some obligate species the hyphae overwinter within twigs or buds, and the mycelium is perennial.

Asexual or Imperfect Stage.—The majority of Ascomycetes have one or more asexual or conidial stages. The conidia of the asexual stage vary greatly, depending on the species. They are one-, two-, or many-celled and are borne free on the ends of conidiophores or in fruiting bodies which, according to their structure, are called *acervuli, pycnidia,* or *sporodochia* (Figs. 8.10, 8.11, 8.12). In some cases conidia are able to survive from year to year in these fruiting bodies.

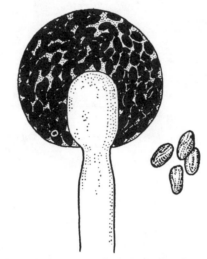

FIG. 8.7. A SPORANGIUM OF THE BREAD MOLD FUNGUS

FIG. 8.8. SCLEROTIA OF *RHIZOCTONIA* ON A POTATO

FIG. 8.9. TAR SPOT OF MAPLE SHOWING BLACK, FLATTENED
STROMATA

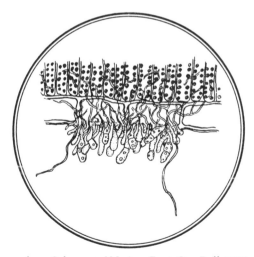

(Redrawn from Schwarze, N.J. Agr. Expt. Sta. Bull. 313)

FIG. 8.10. CROSS SECTION THROUGH AN ACERVULUS, GREATLY
ENLARGED

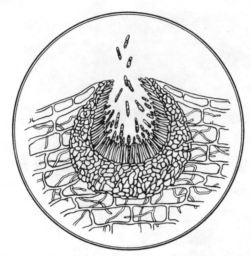

(Redrawn from Schwarze, N.J. Agr. Expt. Sta. Bull. 313)

FIG. 8.11. CROSS SECTION THROUGH A PYCNIDIUM, GREATLY
ENLARGED

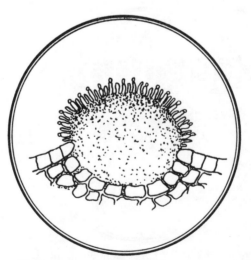

(Redrawn from Heald, Manual of Plant Diseases)

FIG. 8.12. A SPORODOCHIUM

Sexual or Perfect Stage.—The characteristic asci may develop singly or in groups as naked asci (Fig. 8.13) in a palisade-like layer on the surface of the affected tissue, as in the leaf-curl fungi; or they may be grouped into definite fruiting bodies or ascocarps known as *perithecia* (Fig. 8.14) (sphere fungi or Pyrenomycetes), *apothecia* (Fig. 8.15) (disc

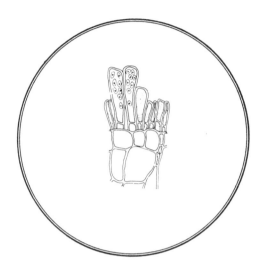

FIG. 8.13. NAKED ASCI

(Redrawn from Melhus and Kent, Elements of Plant Pathology)

FIG. 8.14. CROSS SECTION THROUGH A PERITHECIUM, GREATLY ENLARGED, SHOWING ASCI

or cup fungi, Discomycetes), or *cleistothecia* (Fig. 8.16) (powdery mildews and sooty molds). A few Ascomycetes produce only ascospores, but most produce one or more forms of conidial fruits.

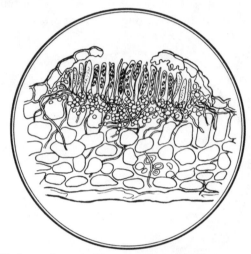

(Redrawn from Heald, Manual of Plant Diseases)

FIG. 8.15. CROSS SECTION THROUGH AN APOTHECIUM, GREATLY ENLARGED

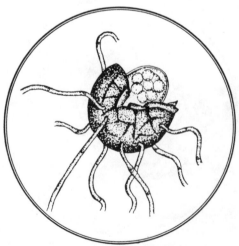

(Redrawn from Schwarze, N.J. Agr. Expt. Sta. Bull. 313)

FIG. 8.16. A CLEISTOTHECIUM, GREATLY ENLARGED

TABLE 8.3

SOME IMPORTANT DISEASES CAUSED BY ASCOMYCETES

Disease	Causal Organism	Host
Leaf curl, leaf blister	*Taphrina* spp.	Peach, nectarine, pear, oak, cherry, plum
Brown rot	*Sclerotinia* spp.	Stone fruits, pome fruits
Anthracnose	*Pseudopeziza ribis*	Currants, gooseberries
Anthracnose	*Glomerella gossypii*	Cotton
Powdery mildews	*Sphaerotheca* spp.	Rose, strawberry, peach, gooseberry
	Erysiphe spp.	Pea clover, cereals, grasses, cucurbits, composites, etc.
	Podosphaera spp.	Apple, pear, quince, cherry, etc.
Black spot	*Diplocarpon rosae*	Rose
Canker	*Nectria* spp.	Apple, pear, boxwood, many other deciduous trees and shrubs
Root, stalk, and ear rot, seedling blight	*Gibberella saubinetii*	Corn, small grains
Ergot	*Claviceps purpurea*	Rye and other cereals
Black knot	*Dibotryon* spp.	Plum, cherry, currant, gooseberry
Bitter rot	*Glomerella cingulata*	Apple, pear, quince, grape
Endothia canker or blight	*Endothia parasitica*	Chestnut
Black rot	*Guignardia bidwellii*	Grape
Leaf spot	*Mycosphaerella* spp.	Strawberry, *Rubus* spp., currants, gooseberries, peas, beets
Scab	*Venturia* spp.	Apple, crabapple, pear, poplar
Dutch elm disease	*Ceratostomella ulmi*	Elm
Sclerotinia wilt and rot diseases	*Sclerotinia sclerotiorum*	Vegetables, field crops, flowers

The Basidiomycetes.—These include the rust, smut, and palisade fungi. They are characterized by a septate mycelium, often with "clamp connections" (Fig. 8.2A) and by a typically club-shaped, nonsegmented basidium on which the spores (basidiospores, usually four) are formed (Fig. 8.17).

Vegetative Structures.—The conspicuous parts of most species of Basidiomycetes are the fruiting bodies (*basidiocarps* or *sporophores*) and the spores, but the actual body of the fungus is the inconspicuous mycelium usually well hidden in the organic matter or in the host's tissues. The mycelium is usually present as felts or wefts of hyphae, as delicate mycelial strands, or as thickened, interlacing strands of mycelium called *rhizomorphs* (Fig. 8.18). The mycelium of parasitic species is intercellular—the rusts have haustoria for penetrating into cells.

The Smuts (Ustilaginales).—The smuts are obligate parasites that are responsible for enormous losses of cereals. Specific smut-producing

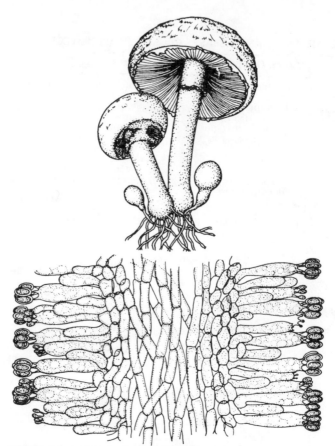

FIG. 8.17. TOADSTOOLS, WITH ENLARGED DRAWINGS OF BASIDIA AND
BASIDIOSPORES

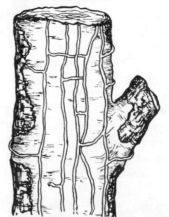

FIG. 8.18. RHIZOMORPHS UNDER THE BARK OF A TREE

fungi generally attack only one organ, but any organ above ground may be attacked by smuts. The smut spores (teliospores) typically have thickened walls, are smooth or sculptured, and are usually black, brown, yellow, or violet. The black spore masses formed by the smuts either break up into a dust-like powder easily blown by the wind (loose smuts) (Fig. 9.19) or remain in place more or less covered (kernel or covered smuts).

Some smut spores may germinate only after a period of dormancy; others may germinate immediately after maturity. They may retain their vitality for several years. In germination they produce a basidium (promycelium) with primary or secondary sporidia (basidiospores) or in some cases many-branched hyphae (Fig. 8.19). The sporidia may infect seedlings before they emerge from the soil, as in oat smut and wheat bunt; they may infect any embryonic tissues, resulting in local lesions in various organs, as in corn smut; or they may infect the ovary through the stigma of the flower and remain as dormant mycelium in the seed until it germinates, when infection becomes systemic, as in loose smut of wheat and naked smut of barley. Sporulation occurs when the host flowers. Smut spores will germinate readily in the compost pile, and the sporidia will continue to bud (like yeast cells) under these conditions until enormous numbers of spores have been produced.

TABLE 8.4

SOME IMPORTANT DISEASES CAUSED BY SMUT FUNGI

Disease	Causal Organism	Host
Loose smut	Ustilago avenae	Oats
	Ustilago tritici	Wheat, rye
	Ustilago nuda and nigra	Barley
Covered or kernel smut	Ustilago levis	Oats
	Ustilago hordei	Barley
	Ustilago fischeri	Corn
	Ustilago cruenta	Sorghum
Boil or common smut	Ustilago maydis	Corn
Bunt or stinking smut	Tilletia caries	Wheat
	Tilletia foetida	Wheat
Black smut	Tilletia horrida	Rice
Onion smut	Urocystis cepulae	Onion
White smut	Entyloma spp.	Spinach, dahlia, sunflower

The Rust Fungi (Uridinales).—Rusts are obligate parasites. The name of the order is derived from the rusty spore pustules or sori that are typical of certain stages. Rusts have been mentioned in connection with plants ever since their cultivation began. They are responsible for great crop losses. The life histories of some rusts are very complex and in some cases are not completely known or understood.

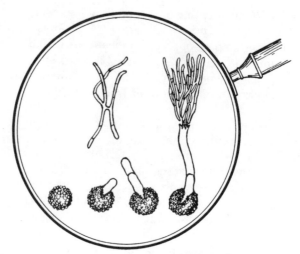

FIG. 8.19. THE GERMINATION OF A SMUT SPORE SHOWING THE FOR-
MATION OF THE BASIDIUM BEARING BASIDIOSPORES (SPORIDIA)

Vegetative Structures.—The myeclium of rusts is generally intercellular, branched, and septate, with haustoria present. Yellowish or orange-red drops of oil may be observed in the mycelium. The mycelium may be annual or perennial depending on the rust species and on the suscept.

Reproductive Structures.—One of the striking features of rusts is their different spore forms (*polymorphism*). Many rusts may produce as many as five different spore forms in the course of their complete life cycle. Another peculiar characteristic of rusts is the necessity in many species for two unrelated hosts to complete their developmental cycle. Such rusts are termed *heteroecious* (as opposed to *autoecious*), with certain spore forms developing on one host and the other spore forms on the wholly unrelated species.

Spore Forms.—The five spore fruits and spore forms in the typical life cycle of a heteroecious rust are listed in Table 8.5 and are shown in Fig. 8.20, but in some species one or more stages in the developmental cycle are regularly missing.

TABLE 8.5

SPORE FRUITS AND SPORES OF RUSTS

Symbol	Spore Fruits	Spores
0	Pycnia	Pycniospores (spermatia)
I	Aecia	Aeciospores
II	Uredinia (sori)	Urediniospores
III	Telia (sori)	Teliospores
IV	Basidia (promycelia)	Basidiospores (sporidia)

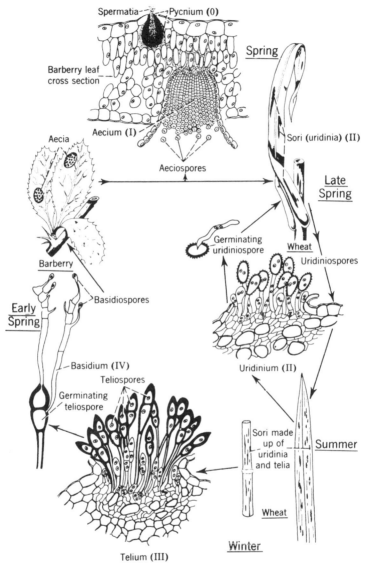

FIG. 8.20. THE LIFE CYCLE OF THE STEM RUST OF WHEAT

If pycnia (0) and aecia (I) are on the same host as telia (III), the rust is autoecious; and if (0) and (I) are on different species than (III), the rust is heteroecious.

The existence of specialized races among rusts, especially those of cereals, is more common than among powdery mildews. Some rust races may be very virulent and may attack a wide number of varieties of the species, whereas others may produce disease in one variety only. Biological specialized races are distinguished by adding a third name (subspecies) to the binomial species. In turn each subspecies is made up of a number of physiological races (designated by a number) which differ in ability to infect various varieties of the same host, e.g., wheat rust, *Puccinia graminis tritici*, has over 200 races to which wheat varieties vary in susceptibility.

TABLE 8.6

SOME IMPORTANT RUST DISEASES

Disease	Causal Organism	Host
Stem rust	*Puccinia graminis tritici*	Wheat, grasses (alt. host, barberry)
	Puccinia graminis avenae	Oat, grasses (alt. host barberry)
	Puccinia graminis secalis	Rye, grasses (alt. host barberry)
	Puccinia graminis hordei	Barley (alt. host, barberry)
Flax rust	*Melampsora lini*	Flaxes
Stripe rust of wheat and rye	*Puccinia glumarum*	Wheat, rye, grasses (alt. host?)
Crown rust of oats	*Puccinia coronata*	Oats, grasses (alt. host, buckthorns)
Cedar-apple rust	*Gymnosporangium juniperi-virginianae*	Apples (alt. host, *Juniperus* spp.)
Blister rusts	*Cronartium* spp.	Five-needled pines (alt. host, currants and gooseberries)
Asparagus rust	*Puccinia asparagi*	Asparagus
Bean rust	*Uromyces phaesoli*	Beans and lima beans
Carnation rust	*Uromyces caryophyllinus*	Carnations
Hollyhock rust	*Puccinia malvacearum*	Hollyhocks, common mallow

The Palisade Fungi (Homobasidiomycetes).—These fungi get their name from the palisade-like arrangement of the basidia. This group is of great importance in relation to diseases of growing trees and in the decays of logs and lumber (Fig. 8.21). A few may attack herbaceous plants. The distinguishing feature of most of this group is the large fruiting bodies, which are often mistaken for the fungous body itself. The fungus is concealed in the host and undergoes considerable growth before it produces the conspicuous fruiting bodies (*sporocarps* or *basidiocarps*, Fig. 8.22).

The Mycelium.—The most distinguishing characteristics of the mycelium of this group are its clamp-like connections, where the septations occur, and the binucleate condition of the hyphal cells. The pres-

FIG. 8.21. A ROT IN A MAPLE CAUSED BY A POOR AND UNPROTECTED
PRUNING CUT

FIG. 8.22. SHELF FUNGI (FOMES IGNIARIUS) ON A BEECH

ence of these clamp connections is often a convenient sign of recognizing a fungus of this group (Fig. 8.2A).

Reproductive Structures.—Basidia are formed from club-shaped hyphae that usually bear four basidiospores, but the number may vary from two to eight. These basidia are arranged side-by-side in a layer and arise in various fleshy, corky, or woody structures known as toadstools (Fig. 8.17), mushrooms, shelf fungi, conks, puffballs, etc. Some of these sporophores open at maturity to expose the basidia, so that the basidiospores may be forcibly released, and others remain closed and are ruptured by natural forces. The presence of these structures on trees, logs, or lumber is of definite diagnostic significance in identifying many woodrot fungi. These fungi, according to their species, may attack the roots of plants and cause root rots; the heartwood of trees and cause heart rots; or the sapwood of trees and cause sapwood rots.

TABLE 8.7

SOME IMPORTANT DISEASES CAUSED BY PALISADE FUNGI

Disease	Causal Organism	Host
Rhizoctonia	*Corticium vagum*	Potato, etc.
Canker	*Septobasidium pedi- cellatum*	Apple, pear, and some forest trees
Seedling blight	*Typhula* spp.	Cereals, beets, cabbage, etc.
Heartwood or sapwood rots	*Hydnum* spp.	Maple, beech, oak, fir
Texas root rot	*Hydnum omnivorum*	Cotton, etc.
Timber rot	*Poria incrassata*	Coniferous trees and lumber
White wood rots	*Fomes* spp.	Various deciduous trees
Red heart rot	*Fomes laricis*	Conifers
White rot	*Polyporus squamosus*	Fruit, nut, and shade trees
Dry rot	*Lenzites sepiaria*	Coniferous timber
Schizo- phyllum rot	*Schizophyllum alneum*	Shade, nut, and fruit trees
White- streaked sapwood rot	*Pleurotus ostreatus*	Maple and deciduous trees
Mushroom root rot	*Armillaria mellea*	Numerous trees
Brown- mottled rot	*Pholiota adiposa*	Deciduous trees and conifers
Sapwood rot	*Polystictus pergamenus*	Deciduous trees

Deuteromycetes (Fungi Imperfecti).—Some 15,000–20,000 species are placed in this class, which is made up of the unplaced forms of the fungous world. It is purely an arbitrary and an unnatural classification in which only the conidial or imperfect stage is known. Generally, as soon as the perfect or sexual stage is found, a fungus belonging to this group is transferred to the correct class. Because of this, many fungi bear two or more synonymous binomial names—a name given to the conidial or imperfect stage (Spermatophyte) and the correct name given to the acigerous or perfect stage (Sporophyte) which was previously undiscovered or whose relationship to the conidial stage was unrecognized. Most of the Fungi Imperfecti probably will be found to belong to the Ascomycetes, but some may belong to the Phycomycetes or Basidiomycetes. Fungi Imperfecti include both parasites and saprophytes, some of which are very destructive to plants. The imperfect stages are represented by conidiophores, pycnidia, acervuli, or sporodochia. Most of the pathogenic Fungi Imperfecti produce symptoms such as fruit and vegetable rots, leaf spots, wilts, and cankers.

Form-Order Mycelia Sterilia.—Those fungi which have no known spore stage are grouped in this order under the Deuteromycetes. It includes two genera of importance in crop production: *Rhizoctonia* spp. which produce damping-off, root rots, and stem cankers of various plants, especially potatoes; and *Sclerotium* spp. producing crown and stem rots of many plants. A basidial saprophytic stage (*Pellicularia filamentosa*) has been found in *R. solani* indicating that it is actually a Basidiomycete. Many of the Mycelia Sterilia form sclerotia as the resting or overwintering structures.

TABLE 8.8

SOME IMPORTANT DISEASES CAUSED BY DEUTEROMYCETES

Disease	Causal Organism	Host
Gray mold or botrytis blight	*Botrytis cinerea*	Buds, fruits, roots, and stem of numerous hosts
Wilt	*Verticillium* spp.	Potato, maple, tomato, etc.
Wilt	*Fusarium* spp.	Cabbage, flax, tomato, potato, cotton, etc.
Leaf spot	*Cercospora beticola*	Beets, etc.
	Cylindrosporium spp.	Cherry and plum
Wilt	*Cephalosporium* spp.	Elm, persimmon, etc.
Anthracnose	*Colletotrichum* spp.	Cucurbits, bean, etc.
Blight	*Coryneum beijerinckii*	Stone fruits
Black leg	*Phoma lingam*	Crucifers
Early blight	*Alternaria solani*	Potato, tomato
Late blight	*Septoria* spp.	Celery
Diplodia disease	*Diplodia* spp.	Corn, citrus
Strumella disease	*Strumella coryneoidea*	Chestnut, oak
Blue mold	*Penicillium* spp.	Apple, pear, citrus

QUESTIONS FOR DISCUSSION
1. How should one proceed to determine the cause of a plant trouble?
2. Which is easier to control, a bacterial disease or a fungous disease?
3. Since plant-attacking bacteria do not form spores, how do the bacteria survive and disseminate themselves?
4. Why must fungi by nature be disease-producing when they live on higher plants?
5. Explain why diseased lesions on foliage are usually round? Why are some angular?
6. In what ways are fungous spores very like the seeds of higher plants?
7. If a sterile nutrient culture medium is exposed to the open air for a short time, it will become contaminated with fungi. What does that show?
8. How are fungous spores disseminated?
9. What does the presence of shelf fungi on a living tree indicate?
10. Why are shelf fungi nearly always present on dead tree trunks or on stumps?
11. Of what importance is it to recognize the reproductive structures of fungi?
12. What are swarm spores and how do they function?
13. What are some of the ways in which fungi overwinter?

Chapter 9

The Annual Life History of Pathogenes

The annual life history of a disease-causing organism consists of all that the organism habitually does, all the stages and cycles that it goes through, and the time spent in each stage from the beginning of its existence (usually in the spring) until one year later. The life history is usually made up of one or more life cycles—initiated by the transfer of some reproductive portion of the organism (*the inoculum*) to a new growing place (*the infection court*) and completed with the cessation of activity therein.

TYPES OF LIFE CYCLES

There may be two kinds of life cycles: primary ones first initiated by the disease-causing organism after a period of seasonal inactivity (usually after the winter rest); and one or more secondary ones initiated by the inoculum from the primary life cycle or other secondary ones. This applies generally to regions where winter and summer or a dry and a wet season occur. For example, in apple scab the ascospores produced in perithecia in the overwintering leaves on the ground are the source of the primary cycles on the new leaves in the spring. Conidia produced in these primary lesions fall on leaves and fruit and produce secondary cycles with more conidia; these in turn produce other cycles, and so on during the summer and fall as long as the weather is favorable (Fig. 9.1).

Some disease-causing organisms under certain environmental conditions may show only secondary cycles. This usually occurs where climatic conditions are favorable for continuous activity of the organism, for example, in greenhouses and in tropical climates. Carnation rust on carnations perpetuates itself only by a succession of secondary cycles, by means of its urediniospores alone.

THE TWO PHASES IN THE LIFE CYCLE OF PATHOGENES

The annual life history of a disease-causing organism may be represented by a circle, segments of which represent the four seasons of the

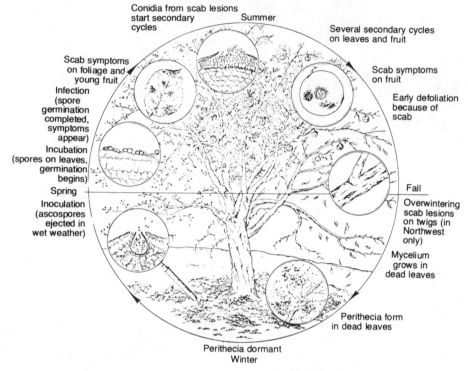

Conidia from scab lesions start secondary cycles

Summer

Several secondary cycles on leaves and fruit

Scab symptoms on foliage and young fruit

Infection (spore germination completed, symptoms appear)

Incubation (spores on leaves, germination begins)

Spring

Inoculation (ascospores ejected in wet weather)

Scab symptoms on fruit

Early defoliation because of scab

Fall

Overwintering scab lesions on twigs (in Northwest only)

Mycelium grows in dead leaves

Perithecia form in dead leaves

Perithecia dormant
Winter

FIG. 9.1. THE ANNUAL LIFE CYCLE OF THE APPLE SCAB FUNGUS

year. A line drawn through the circle separates the two phases usually exhibited during the life history of a disease-causing organism. The disease-producing phase (*pathogenesis*) occurs during the growing season, whereas the development in dead material or dormant phase (*saprogenesis*) usually occurs during the cold or (in the tropics) dry season. During pathogenesis the pathogene is usually actively attacking the tissues of its host and producing characteristic symptoms of disease; during its inactive stage the disease-causing organism may be dormant or developing in the dead tissues of its host as a saprophyte (Fig. 9.1).

THE THREE STAGES OF DISEASE ATTACK

During pathogenesis, a pathogene may pass through one or more life cycles, each of which exhibits three rather distinct successive stages: *the inoculation stage, the incubation stage,* and *the infection stage.* This closely parallels the life cycle of human and animal diseases, which also show these three stages. The length of each of these varies with the nature of the disease-causing organism and that of the host plant, as well as with the environmental conditions involved. Secondary life cycles are represented in the figures by a small circle in which the three stages are repeated as often as a new cycle occurs (Fig. 9.1).

The Inoculation Stage

This stage begins at the moment that the inoculum leaves its source and ends when it lands on or in the infection court (Fig. 9.1). Inoculation is the act of transfer of the inoculum from its source to the infection court. The source of the inoculum is generally an infected organ or a lesion on or from some part of a suscept, such as a canker or a diseased leaf or fruit. The soil is also very often a source of the inoculum. An infection court may be a wound or break in the bark or epidermis, water pore, stoma, leaf surface, stigma, flower nectaries, lenticels, etc. on a susceptible plant.

Generally the inoculum is transferred from its source by some agency other than itself, referred to as the *inoculating agent* or vector (e.g., insects transmitting diseases). The usual inoculating or disseminating agents are air currents, wind, rain, splashing or running water, insects, and many other animal organisms including man. Each disease-producing organism may make use of one or more of these agents for its dissemination.

Liberated inoculum may be short-lived and fragile, or long-lived and resistant. If short-lived and fragile, it may survive air transport for only a few hours or only a short distance from its source. For example, zoöspores and sporangia of late blight of potatoes under dry atmospheric conditions; and sporidia of white pine blister rust on *Ribes* spp. and of cedar-apple rust on *Juniperus* spp. will not survive air transport for more than about ¼ mile and 3 miles respectively. If long-lived and resistant, it can withstand dessication, low temperatures and high light intensity. Urediniospores of stem rust of wheat, for example, are carried by winds at high altitudes over great distances for long periods of time and still retain viability.

One of the common methods for inoculum to be introduced into plants is through wounds made by insects in feeding. Many viral and some bacterial and fungous diseases are known to be transmitted by insects, and in some cases insects are the only known inoculating agents.

The Incubation Stage

The incubation stage starts from the moment that the inoculum lands on the infection court and ends with the beginning of the reaction of the plant to the irritation of the disease-producing organism (Fig. 9.1). The incubation period includes an inactive period or periods, development and germination of the spores, and the penetration (in some cases also the multiplication and spread) of the disease-causing organisms. During this stage the pathogene prepares to attack the living cells or tissues of its host.

The duration of the incubation stage is generally measured from the time when conditions were favorable for the inoculation to take place until the first evidences of disease in the host plant can be detected. This is not an accurate measure: it is not possible to know exactly when the inoculum leaves its source; and the first noticeable reactions of the plant to the invasion of the pathogene occur some time later than actual irritation began. However, this measure is the only practical one to date.

Survival of Inoculum.—When the proper environmental conditions are not present for the inoculum to begin activity, it either dies or is able to survive in a quiescent condition for a variable period until favorable conditions arise. Spores vary greatly in their ability to survive conditions unfavorable for incubation. Conidia are able to survive for a few days to several weeks; other spores, such as chlamydospores of smuts, may survive for a year or many years and then germinate under the right conditions. When conditions become unfavorable for further development during the incubation stage, the inoculum may perish, or its activity is temporarily halted, to resume again later when conditions become favorable.

Activity of the Inoculum.—In most cases the inoculum begins activity as soon as it lands on the infection court, since the damp or wet conditions necessary for inoculation usually carry over into the incubation period. During the incubation period spores germinate by sending out an infection thread, in contrast to the inoculum of bacteria and viruses that must gain entrance into the plant tissues to start incubation, multiplication, and spread. The conidia of some powdery mildews germinate best in fairly dry conditions. Some fungous spores will germinate regardless of where they land, provided they have the right temperature and moisture conditions; but others require in addition a chemical stimulus from the tissues of their host before germination will occur.

Most disease-causing organisms must enter the tissues of the host to produce disease. Some are capable of penetrating directly through the uninjured cuticle and epidermis; others enter only through natural openings; and still others can enter only through wounds or lesions. Ingress of fungous pathogenes into host tissues is the final act of incubation. It is this final act that most protective fungicides tend to prevent.

The Infection Stage

This stage includes that portion of disease activity during which the disease-causing organism is irritating the plant and producing symptoms characteristic of this irritation (Figs. 7.13, 9.1). This stage begins within a few days, and ends with the final reaction (usually death) of the plant or plant part to the activities of the pathogene. The infection stage may be short or long, often extending for years before the plant succumbs from the effects. Plant recovery from a disease is usually rare unless infection is confined to the leaves or fruit of trees or shrubs. The period of infection may last from a few days to several years, depending on the nature of the pathogene, age and type of host tissues, temperature, moisture, and possibly other factors. Brown rot on stone fruits destroys all living cells within a few days, but certain heart rots of trees continue their infection for years without destroying the trees. Potato mosaic infection may extend endlessly through the medium of the tubers.

During the active period of infection, the pathogene is growing, spreading, and multiplying; and the suscept shows primary and secondary symptoms. Then activity may gradually cease, and secondary decay or healing may set in. In many cases secondary decay or healing goes on during active infection.

Progress of Infection.—Infection may be continuous or discontinuous. Infection in some diseases may be stopped permanently or temporarily by unfavorable temperature, water supply, host activity, etc., and may continue again when conditions become favorable. Many canker diseases (such as fire blight and black rot of apple) and some leaf diseases (such as alternaria blight of potatoes and frog-eye (black rot) on apple leaves) show zonations due to alternate halting and renewing of infection activities. Often infection is continuous and ends only with the death of the plant or plant organ invaded.

Infection Areas.—Many organisms causing plant diseases show a definite preference for or restriction to certain plant tissues or organs. A number of bacterial and fungous pathogenes invade the xylem vessels and cause wilts of various plants. Some are limited to sapwood or heartwood, causing wood rots, but the majority confine themselves to the phloem and to parenchymous tissues such as the cortex, and to leaf and fruit parenchyma, causing cankers, leaf spots, rots, galls, etc.

Nature of the Attack.—Disease-producing organisms may injure plants in a number of ways:

1. They may affect cells or tissues by means of toxins that they produce, as in the case of certain plant wilts and the Dutch elm disease. These toxins may produce overstimulation, inhibition of growth, or death of cells.
2. They may rob cells or tissues of food through their haustoria or by enzymic action within the plant.

3. They may penetrate into cells and destroy their cytoplasmic membranes. The earliest symptom of many diseases is a water-soaked appearance of the tissues, indicating a breakdown of the cell membranes.
4. They may make it easy for secondary pathogenes to gain entrance and break down tissues further, as in the case of numerous rots.

ENVIRONMENTAL CONDITIONS IN RELATION TO INFECTIOUS DISEASE OUTBREAKS

Conditions Necessary for an Enphytotic

When a destructive plant disease sweeps over an area, it is referred to as an epiphytotic disease. Epiphytotic diseases may be local, involving only small areas, or they may be widespread, involving large areas, states, and even continents. Three conditions are necessary for a disease to become epiphytotic: (1) an abundance of inoculum; (2) concentrations of host plants in a susceptible stage of growth; and (3) favorable environmental conditions for the initiation and completion of the inoculation, incubation, and infection stages of the pathogene's primary and secondary life cycles. Inoculum for an enphytotic may originate from a few infected plants, from infected seed and lack of crop rotation, from overwintering spore stages, or from secondary life cycles. The amount of injury to any crop is governed by the degree of variation of the three conditions above from the optimum for each particular disease.

Many diseases appear nearly every year with little damage to crops. These are termed *sporadic diseases,* but occasionally they may become epiphytotic under favorable conditions.

Environmental Factors and Enphytotics

Environment, by its effects on the disease-producing organism, on the inoculating agents or vectors, and on the host plants, either favors or does not favor the development of a disease. The environmental factors involved are: (1) soil conditions (physical, chemical, and biological composition), water content, availability of nutrients, and chemical reaction (pH); and (2) atmospheric conditions (temperature, moisture, sunlight, and wind). All these factors interact with one another to affect both the disease-producing organism and the suscept at the same time. However, factors favorable for the development of the disease-producing organism are not always optimum for host susceptibility, and vice versa. Each disease and each stage of a disease has an ideal range for each environmental factor, and any departure from this range tends to reduce the prevalence of the disease.

Environment not only influences the growth of plants but also causes them to vary in their susceptibility to disease invasion. A plant may be very susceptible under one set of environmental conditions and highly resistant under another. Enphytotics occur locally in many areas every year, and generally control or preventive practices are designed to alter one or more of the three predisposing factors mentioned previously. Some of the most destructive enphytotics have been caused by disease-producing organisms new to a particular area and to a highly susceptible host. The chestnut blight and the Dutch elm disease are examples of such enphytotics.

Temperature and Moisture.—These are two of the most important interacting factors regulating disease outbreaks. The majority of bacterial and fungous disease-producing organisms require moisture or damp conditions for definite periods within a definite temperature range for the liberation of the inoculum and the completion of the incubation stage on the infection court (Fig. 9.2). Some of the powdery mildews, however, can complete incubation under dry conditions.

Optimum air or soil temperatures vary greatly with different disease-producing organisms (Table 9.1) and even with the different stages that make up the life cycle. For example, the organism causing late blight of potatoes requires wet foliage and cool air temperatures for optimum inoculation and incubation (Fig. 9.3) but humid, warm temperatures for optimum infection progress.

In general, it may be stated that diseases tend to become epiphytotic in seasons of the year marked by abundant rainfall, long wet periods, or fogs, as long as temperatures remain within the effective range of development of the disease-producing organisms. In wet, rainy springs and summers, numerous diseases may be expected to show up on plants, whereas in dry springs and summers most diseases are scarce or practically nonexistent. Many of our destructive plant disease-producing organisms, such as those that cause fusarium wilts, onion smut, rhizoctonia of potatoes, potato scab, and carotovorus soft rot, may survive several years in the soil under the proper temperature and moisture conditions.

Soil Reaction and Infectious Diseases.—Disease-producing organisms vary greatly in their requirements of the pH of soil or plant juices. Diseases, such as cotton wilt, tomato wilt, rhizoctonia root rot, and the club root of crucifers, are favored by a low pH or acid soil reaction; whereas Texas root rot, potato scab, black root rot of tobacco, and soil rot of sweet potatoes are favored by a high pH or alkaline soil reaction. These diseases are often controlled by altering the soil reaction, but what actually takes place in the plant or soil to make conditions

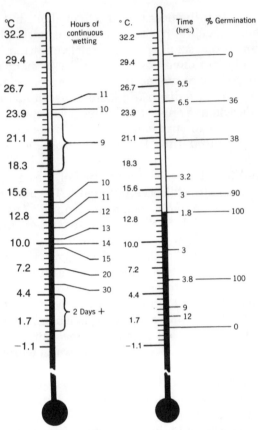

(Redrawn from Mills and Dewey, Cornell Ext. Bull. 711)

FIG. 9.2. HOURS OF CONTINUOUS WETNESS NECESSARY AT VARIOUS
TEMPERATURES TO COMPLETE THE INOCULATION AND INCUBATION
STAGE OF THE APPLE SCAB FUNGUS IN THE SPRING

(Based on data from Melhus and Kent,
Elements of Plant Pathology)

FIG. 9.3. RELATION BETWEEN TIME, TEMPERATURE, AND PER CENT
GERMINATION OF POTATO LATE BLIGHT CONIDIA UNDER OPTIMUM
MOISTURE CONDITIONS

unfavorable for the disease is often not clearly understood. Varietal
resistance to disease may in many cases be due to unfavorable chemical
reaction for the disease-producing organism within the plant tissues.

TABLE 9.1

OPTIMUM INCUBATION TEMPERATURES OF SOME DISEASES UNDER THE PROPER
MOISTURE CONDITIONS (DATA FROM VARIOUS SOURCES)

Disease	Air Temperatures, °	
	° F	° C
Apple scab	65–75°	18–24°
Brown rot	high	
Cedar-apple rust	65°	18°
Late blight of potato	52.6°	11°
Peach leaf curl	low	
Physoderma disease of corn	82–84°	28–29°
Disease	Soil Temperatures, °	
Fusarium wilt of tomato	80.6°	16–18°
Bunt of wheat	60–64°	10–12°
Onion smut	50–54°	16–29°
Cabbage yellows	79–84°	
Watermelon wilt	78–82°	26–28°
Potato rhizoctonia	59–70°	15–21°

THE LIFE HISTORIES OF SOME DISEASE ORGANISMS

Late Blight of Potatoes (Caused by Phytophthora infestans)

Host Plants.—Potatoes, tomatoes.

Symptoms.—Waterlogged dark spots that spread rapidly, killing the leaves. In dry weather the leaves turn dark brown, but in damp weather the leaves remain dark with white groups of conidiophores and conidia covering the affected areas (Fig. 7.1). Stems, tubers, and fruits may also be attacked, resulting in dry or wet rots depending on the humidity (Fig. 9.4).

Life History.—See Figs. 9.5 and 9.6. Two races are present in the United States; only one is virulent enough to attack tomatoes. Repeatedly growing the pathogene on highly resistant varieties has been shown to increase its virulence so that resistant varieties may be attacked severely. The disease is usually brought to a field in infected seed pieces, but airborne spores are responsible for enphytotics.

Fire Blight (Caused by Erwinia amylovora)

Host Plants.—Pear, apple, quince, and many other plants of the family Rosaceae.

Symptoms.—The organism thrives in succulent, vigorous-growing tissues. During its development it may cause such symptoms as leaf blight, fruit blight, blossom blight, twig blight, and branch and trunk cankers. The dead leaves and fruit cling to the branches and do not fall off readily (Fig. 9.7).

Life History.—See Fig. 9.8. Inoculation takes place mainly through

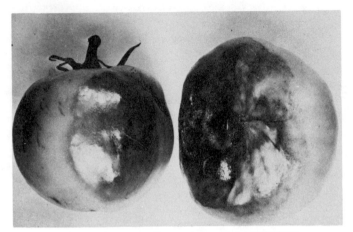

FIG. 9.4. LATE BLIGHT PRODUCES A ROT IN TOMATOES

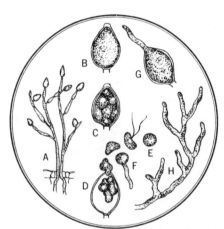

(Redrawn from Melhus and Kent,
Elements of Plant Pathology)

FIG. 9.5. THE LATE BLIGHT FUNGUS OF POTATOES

A, Conidiophores bearing conidia; B, C, D, Zoöspore formation
from a conidium; E, F, Resting and germinating zoöspore; G,
Direct germination of a conidium; H, Mycelium of fungus.

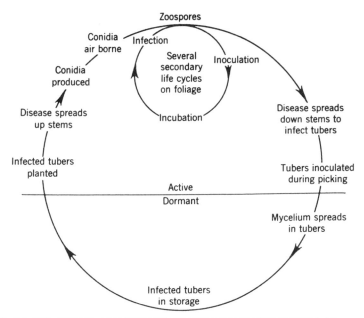

FIG. 9.6. THE ANNUAL LIFE HISTORY OF THE LATE BLIGHT FUNGUS OF POTATOES

nectaries, stomata, lenticels, and wounds. The dissemination of bacteria is mainly by insects and rain.

Damping-off (Caused by *Pythium debaryanum*)

Host Plants.—Seedlings of most plants, especially those grown in greenhouses, coldframes, and hotbeds.

Symptoms.—Damping-off may be caused by several other pathogenes, but the one above is the most important. In pre-emergence damping-off the seeds or seedlings rot before emergence takes place. In post-emergence damping-off the seedlings are attacked at the soil line or below; they wilt and then fall over.

Life History.—See Figs. 9.9 and 9.10.

Cotton Wilt (Caused by *Fusarium vasinfectum*)

Host Plants.—Cotton.

Symptoms.—Premature wilting and death of plants. Brown or black discoloration of the vascular bundles appears in the roots and stem. Symptoms are most evident when plants are 8–12 in. high, but the same fungus may also cause a seedling rot and early wilt similar to damping-off. The disease is most prevalent in light, sandy soils.

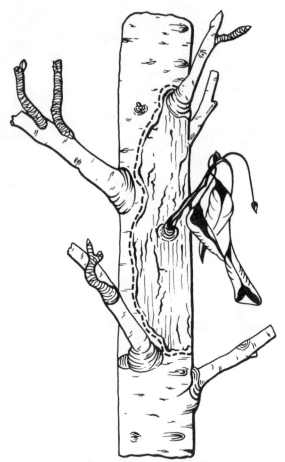

(After Hildebrand, Cornell Ext. Bull. 405)

FIG. 9.7. A FIRE BLIGHT CANKER IN THE DORMANT SEASON

Dotted line indicates location of bacteria.

Life History.—See Fig. 9.11. Disease is disseminated by implements, animals, water, and wind.

Soft Rot of Vegetables (Caused by *Erwinia carotovora*)

Host Plants.—Most root crops and some other vegetables such as lettuce, cabbage, celery, and tomatoes.

Symptoms.—Produces a soft, slimy rot of the infected tissues, usually roots. Epidermis may remain intact with the rot internal. Rot produces a very foul odor and may develop in storage as well as in the field, under warm, damp conditions.

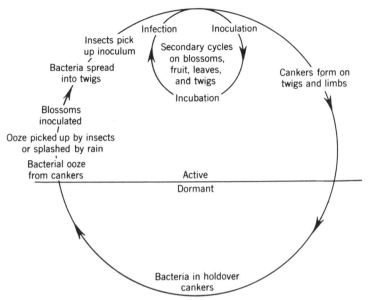

FIG. 9.8. THE LIFE HISTORY OF THE FIRE BLIGHT BACTERIUM *(Erwinia amylovora)*

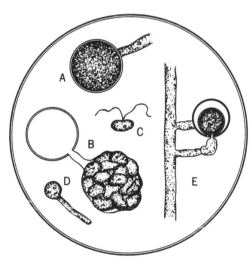

*(Redrawn from Melhus and Kent,
Elements of Plant Pathology)*

FIG. 9.9. THE REPRODUCTIVE STRUCTURES OF THE FUNGUS CAUSING
DAMPING-OFF

A, Conidium; B, Germinating conidium; C, Zoöspore; D, Germinating zoöspore; E, The formation of an oöspore.

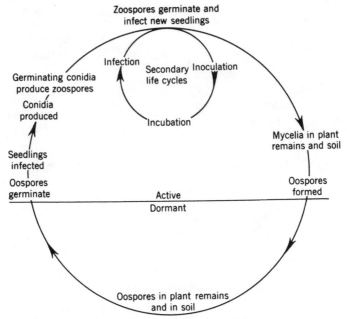

FIG. 9.10. THE LIFE HISTORY OF THE DAMPING-OFF FUNGUS *(Pythium debaryanum)*

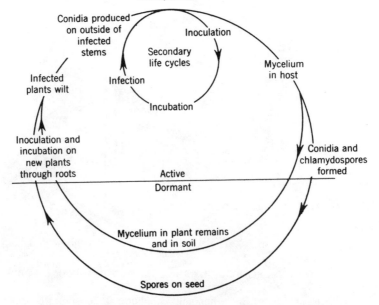

FIG. 9.11. THE LIFE HISTORY OF THE COTTON WILT FUNGUS *(Fusarium vasinfectum)*

Life History.—See Fig. 9.12. The bacteria are easily spread by contact, tools, insects, contaminated manure, etc.

Brown Rot of Stone Fruits (Caused by *Monilinia fructicola*)

Host Plants.—Peach, plum, cherry, apple, and apricot.

Symptoms.—The brown rot fungus may attack the flowers and leaves, producing a blight; the twigs, producing cankers; and the fruit, producing a brown rot. Light-brown, tufted conidiophores usually cover the necrotic lesions on flowers and fruits (Fig. 7.6). Blossom blight occurs under hot, damp conditions. The disease organism usually travels down the flower petiole and enters the stem, producing a girdling canker that wilts and kills the growth above. Fruit is most susceptible when it is ripe and has lesions caused by the plum curculio or scab.

Life History. See Figs. 9.13 and 9.14.

White Heart Rot of Hardwoods (Caused by *Fomes igniarius*)

Host Plants.—Many deciduous trees, especially hardwoods such as oak, maple, birch, willow, apple, butternut, and aspen.

Symptoms.—Fungus attacks mainly the heartwood of the trunks of growing trees as well as of dead trees. Discolored, irregular, yellowish-

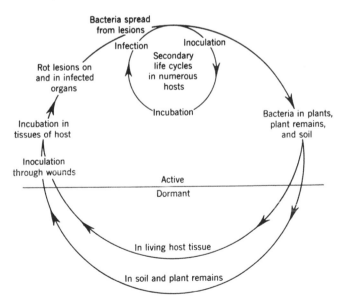

FIG. 9.12. LIFE HISTORY OF THE SOFT ROT BACTERIA *(Erwinia carotovora)*

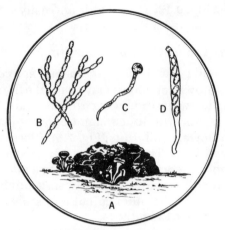

FIG. 9.13. BROWN ROT OF STONE FRUITS
A, A plum mummy with apothecia; B, Conidia; C, Conidium
germinating; D, An ascus.

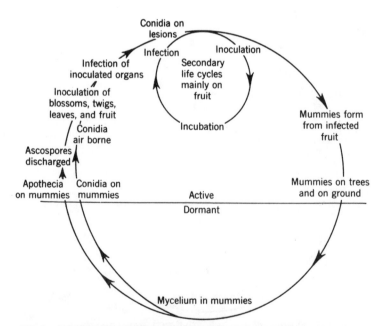

FIG. 9.14. THE LIFE HISTORY OF THE BROWN ROT FUNGUS *(Monilinia fructicola)*
ON PEACH

white areas bounded by dark bands characterize the invaded wood. The fungus sometimes invades and destroys the sapwood and bark of trees. *Life History.*—See Figs. 8.22 and 9.15.

Black Rot of Apple (Caused by *Physalospora cydoniae*)

Host Plants.—Apple, peach, quince, currant, pear, and other woody species.

Symptoms.—Fungus may attack leaves, limbs, and fruit. Brown leaf spots usually of a zonate character and referred to as "frog-eye" are produced. The bark of limbs may be roughened or killed and cracked. The limbs are killed when they are girdled by the cankers (Fig. 9.16). Rot occurs mainly on mature fruit as dark, often zonate, lesions that spread throughout the whole fruit. Fruit turns into a black, shiny mummy with numerous black pycnidia covering it.

Life History.—See Fig. 9.17.

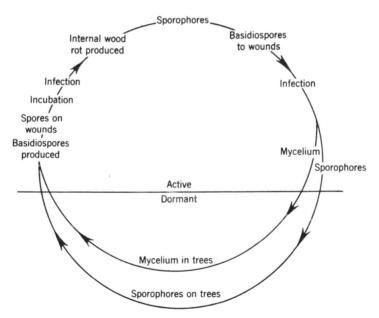

FIG. 9.15. THE LIFE HISTORY OF THE WHITE HEART ROT FUNGUS, *FOMES IGNIARIUS*

FIG. 9.16. BLACK ROT CANKER ON APPLE
Note also the spots caused by the same fungus on the foliage.

Crown Gall (Caused by *Agrobacterium tumefaciens*)

Host Plants.—Most fruits including pome fruits, stone fruits, blackberries, raspberries, loganberries, currants, gooseberries, grapes; also roses, nut trees, numerous woody and herbaceous ornamentals, alfalfa, cotton, beets, turnips, and hops.

Symptoms.—This bacterial organism causes tumor-like, rounded, rough enlargements at the crown, roots, trunks, or branches of susceptible plants (Fig. 9.18). Since the organism can undergo incubation and infection only in wounds it is often found at the point of union of root-grafted stock, but it should not be confused with graft knots, which are overgrowths of excess callus tissue. These galls often interfere with conduction of water upwards in the plant. Young galls are soft or spongy and creamy white; as they enlarge they become harder and turn dark brown.

Life History.—See Fig. 9.18.

Loose Smut of Wheat (Caused by *Ustilago tritici*)

Host Plants.—Wheat, rye, barley.

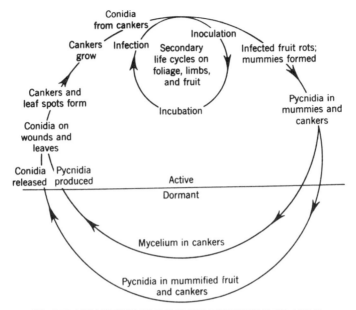

FIG. 9.17. LIFE HISTORY OF THE BLACK ROT FUNGUS ON APPLE

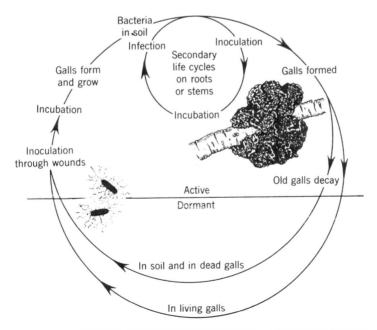

FIG. 9.18. SEASONAL LIFE CYCLE OF CROWN GALL—A BACTERIAL DISEASE

Symptoms.—Black smut masses cover the flower heads of the plants (Fig. 9.19). These spore masses tend to be broken up, and the spores are distributed by the wind. Smutted heads appear early, just as the normal heads are flowering.

Life History.—Smut spores are carried by the wind to flower stigmas where they undergo incubation and infect the ovary. The infection remains dormant in the ovary until the seed that has formed has germinated. Hyphae then travel upward to infect the young spikelets (Fig. 9.20).

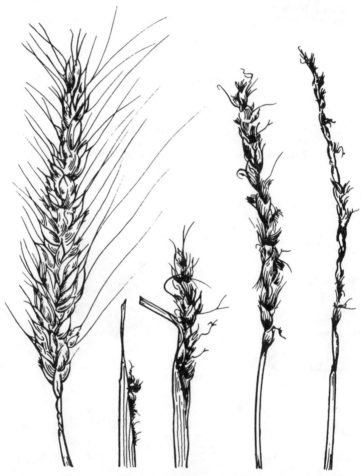

(Redrawn from U.S. Dept. Interior Leaflet 3)

FIG. 9.19. LOOSE SMUT OF WHEAT, THE SPIKE ON THE LEFT IS NORMAL

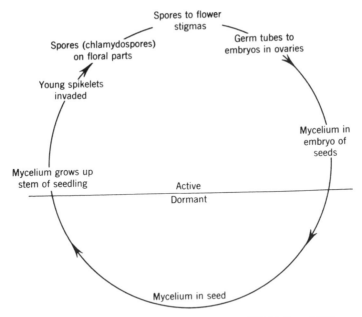

FIG. 9.20. LIFE HISTORY OF LOOSE SMUT OF WHEAT *(Ustilago tritici)*

QUESTIONS FOR DISCUSSION

1. Is a knowledge of the life history of a disease-producing organism necessary for its control?
2. What limits the number of secondary life cycles a disease-causing organism may undergo?
3. Describe the inoculation, incubation, and infection stages of some human diseases.
4. Why did people at one time think that diseases arose spontaneously?
5. What conditions favor the inoculation stage of plant diseases?
6. Do these same conditions favor the completion of the incubation stage?
7. What diseases are most likely to become destructive: those in which the inoculum is air-borne, or those in which the inoculum is water- or animal-borne?
8. Why do diseases often seem to appear out of nowhere under favorable conditions?
9. Will disease-causing spores germinate wherever they may be under favorable weather conditions?

10. Why is it too late to apply a protective fungicide after incubation has been completed?
11. What types of diseases may plants recover from?
12. During what stages are plant disease-causing organisms most susceptible to control?
13. What are the conditions that usually lead up to an enphytotic?
14. Do individual plants vary in their susceptibility to disease attack? Varieties of the same species?
15. What effects do growing conditions have on the prevalence of a disease?
16. Do all disease-producing organisms have the same requirements for optimum development?
17. Why do retarded germinating conditions favor poor germination?
18. Why is crop rotation important in the control of fusarium diseases?
19. Which plants are more subject to disease attack, those growing vigorously or those doing poorly?
20. Would it be possible to eradicate a wind-borne pathogene from a given area?
21. Why is good potato land generally quite acid although potatoes do as well or better at a higher pH?
22. Why do potato or tomato late blight enphytotics usually occur late in the growing season?
23. Why isn't the stem rust of wheat wholly eliminated by the destruction of barberry plants in infested areas?
24. Study each of the life histories presented in this chapter and pick out the possible points where each organism may be attacked.
25. Is correct timing of control measures more important in insect or disease control?
26. Why are chemical measures for disease control best made before rainy periods?
27. Why is fire blight difficult to control once it gets started in an orchard?
28. If apple scab overwinters mainly in the dead leaves, why is the destruction of the dead leaves not a standard practice in commercial orchards?
29. Is this statement correct: most organisms causing rot disease can get started only in lesions?
30. Does a tree's condition have any effect on its susceptibility to wood rots? Cankers?
31. What disease-causing organisms may be carried from plant to plant by means of pruning tools?

Chapter 10

Viruses and Physiopaths as the Cause of Plant Diseases

VIRUSES AS CAUSES OF PLANT DISEASE

Viruses are the causes of many harmful and often fatal human diseases such as smallpox, typhus fever, measles, mumps, and rabies. Plants also exhibit an imposing and growing number of infectious diseases caused by viruses (more than 600) but these are not known to be transmissible to animals. The term "degenerative diseases" has often been applied to plant diseases caused by viruses because these diseases tend to produce a gradual but progressive loss of vigor.

The Nature of Viruses

The nature of the entities causing viral diseases were long a mystery because they could neither be seen through the most powerful microscope, nor filtered by the finest bacterial filters. The electron microscope has revealed a new world of submicroscopic objects that are considered to be the viral particles or bodies themselves.

The smallest bacteria, which measure about $1 \times 1\frac{1}{2}\mu$, are 10–100 times larger than viral particles (virions). They may be of several forms: long and short rods with or without rounded ends and separate or clustered spherical bodies. They have characteristics that place them in both the living and in the nonliving categories.

The characteristics that place them in the living category are: (1) their ability to multiply in cells or tissues after inoculation and produce disease symptoms after a more or less constant elapsed time; (2) their dependence on living cells for growth and reproduction; (3) their fixed thermal death point, which varies with the virus; (4) their ability to form many distinct strains; (5) their selective host ranges; and (6) the presence of RNA or DNA. Usually a given virus tends to attack only plants in one family, but some, such as curly top virus of beets, have been found to attack 220 host species in 41 families; and some, such as corn mosaic virus, have been found only in corn and sorghum.

The characteristics that place them in the nonliving category are their ability to act like lifeless chemicals in the test tube and to be crystallized out of solution, at least in some cases, in the form of a highly infectious crystalline protein. This has led to the fact that viruses contain nucleoproteins, ribonucleic acid (RNA) or deoxyribonucleic acid (DNA), capable of multiplying when they are in living tissues and of acting like living organisms, but acting like lifeless chemicals when they are separated from their living hosts.

The Classification of Viruses

The classification of viruses is controversial with no standard scheme established to date. At present viruses are identified by their common names but the Plant Virus Subcommittee of the International Committee on Nomenclature of Viruses has suggested more specific identification by the use of cryptograms (descriptive symbols) after each common name: e.g. cucumber mosaic virus (R/1: 1/18: S/S: S/Ap). Key: *first element*, type of nucleic acid—R for RNA, D for DNA—and number of nucleic acid strands—1-single, 2-double; *second element*, molecular weight of nucleic acid in millions and the nucleic acid fraction of the viral particle in percent; *third element*, outline of viral particle and protein coat—S spherical, E elongated, U elongated with rounded ends; *fourth element*, host—S seed plant and vector, Ap aphid, Ne nematode, Fu fungus, Th thrips, Cl beetle, Au leafhopper, O spreads without a vector.

Symptoms of Viral Disease

Viral diseases are usually systemic but some may remain localized. They are much more common in herbaceous plants than in woody plants. Viruses usually do not kill plants outright but produce a slow degeneration with stunting and death of tissues as the final symptoms. Some viral diseases are so mild that the plants may bear a crop although it is usually inferior in quality and quantity. In some cases apparent recovery has been noted, but it may be more of a case of resistance or masking of symptoms, as such plants still have the virus in their cells. It has been noted that weather and growth conditions affect the severity of the symptoms. Generally symptoms are most severe in cool weather and under poor growing conditions and are less severe or even masked under warm, vigorous growing conditions. Symptoms are usually hypoplastic, but many virus diseases also produce hyperplastic chlorotic or necrotic symptoms.

Viruses may produce one or more of the following disease symptoms in plants:

Paling. General paling of leaves and other green parts as exemplified by aster yellows, peach yellows, raspberry yellows, tomato yellows (Fig. 10.1).

FIG. 10.1. RASPBERRY YELLOWS
Note pale, undersized leaves.

Mottling of foliage or fruit caused by irregular, alternate patches or spots of light green or yellow and normal green. Puckering and wrinkling and stunting of growth often accompany this symptom which is characteristic of a number of virus diseases known as mosaics (Fig. 7.9). Mottling of the petals of sweet peas and petunias is caused by mosaics, whereas breaking of tulips (production of variegated flowers by solid color varieties) is caused by a virus infection that is perpetuated in some commercial varieties.

Stunting and shortening of internodes (rosetting). One or both of these symptoms may commonly accompany the previous symptoms. Examples are rosette of peach, dahlia stunt (Fig. 10.2), little peach, dwarf of bramble fruits, and witches' broom effect on various plants.

Rolling and curling of leaves as found in leaf roll of grape, curly top of beets, potato leaf roll, and leaf curl of raspberries.

Distortion and malformation of leaves as shown by puckered, twisted, or narrowed leaf surfaces. Examples are shoe-string and fern leaf of tomato (Fig. 10.3), delphinium stunt, and pepper mosaic.

Ring spot. Yellowish rings with necrotic or chlorotic centers on the green background of the leaves, as found in ring spot of dahlias, peonies, tobacco, and lupine mosaic.

FIG. 10.2. DAHLIA STUNT CAUSED BY A VIRUS

Compare with normal plants in the background.

FIG. 10.3. SHOE-STRING SYMPTOMS ON TOMATO INFECTED WITH
CUCUMBER MOSAIC

Necrosis. The browning and death of tissues is a characteristic effect of some viruses and may involve internal or external tissues. External symptoms may be spots, streaks, or canker-like lesions, as in citrus psorosis. Internal symptoms may show a necrosis of the phloem as in phloem necrosis of elm, curly top of sugar beet, and net necrosis of potato.

Some viral diseases are actually the result of attack by two or more viruses at the same time with symptoms unlike those caused by either one alone. For example, streak of tomato is due to the latent virus of potato combined with tobacco mosaic, and rugose mosaic symptoms on potato are caused by a combination of two viruses attacking potatoes.

Transmission of Virus Diseases

Virus diseases may be transmitted by any one or more of the following means.

Mechanical Means.—This involves the actual application of the virus-containing sap to the suscept and is a common method of spreading the more highly infectious viruses such as tobacco, tomato, and cucumber mosaics. First diseased plants, then healthy ones are touched by pruning tools, grafting tools, or the hands. Tobacco smokers or chewers who smoke or expectorate near them may infect healthy plants.

Insects or Other Organisms.—Insect vectors are responsible for the spread of the majority of virus diseases under natural conditions. In some cases the insects' mouthparts merely transfer the inoculum mechanically from diseased to healthy plants, and in other cases the insect appears to be an intermediate host, as in the streak disease of maize and in aster yellows. In these two diseases the virus passes into the insect's digestive tract, and only after incubating for days or weeks and after multiplying can it be supplied to healthy plants through the saliva injected during feeding. Thereafter the insect and its progeny are continuous inoculating agents every time they feed. Arthropods most often responsible for transmitting viral diseases include sucking types such as aphids, leafhoppers, plant bugs, white flies, mealy bugs, mites, thrips; and chewing types such as grasshoppers, flea beetles, and cucumber beetles. Some nematodes may also transmit viral diseases. Sometimes only a certain species of insect is able to transmit a definite virus under natural conditions. The plum leafhopper is the only insect that transmits peach yellows—grafting is the only other way in which it can be transmitted.

Vegetative Propagation.—The use of grafting or budding stock from diseased plants is a common method of transmitting most viruses to healthy plants and frequently no other way of spreading the disease is known. Root contact may spread viruses among adjacent trees. This occurs in the scaly bark disease of citrus and in phloem necrosis of elm. Cuttings, bulbs, tubers, corms, and any vegetative part from a virus-diseased plant will generally establish the disease in the new plants.

This problem may be avoided in some cases by the propagation of virus-free stock even though the "mother" plants are diseased. This is accomplished by meristem culture, tip culture or meristem-tip culture. Essentially, these techniques consist of excising virus-free apical meristem tissues a few millimeters long, growing them aseptically in a culture medium, and transplanting the plantlets into soil to grow to maturity.

Seed.—Most viruses do not enter the seeds of plants attacked; however, some virus diseases such as the mosaics of legumes, lettuce, and petunia are known to be carried in the seed of infected plants and may gain a foothold in a garden by such means.

Soil.—The only viruses known to be transmitted in the soil are wheat mosaic (which survives long enough to affect the wheat crops following on that soil) and big vein of lettuce. The tobacco mosaic virus may be transmitted through infected tobacco refuse in the soil.

Pollen.—Bean mosaic is the only known virus disease of economic importance that has been reported to be transferred to healthy bean plants through the medium of pollen from infected plants.

Dodder.—Dodder is a parasitic vine that forms living bridges from plant to plant and thus transmits the virus.

For the life histories of two typical viruses see Figs. 10.4 and 10.5.

Plant Viroids

Plant viroids are a recently identified class of subviral plant pathogenes composed of naked RNA (without a protective protein coat) and much smaller (about 30 times) than the virions (viral particles) of known viruses. No resting phase has been observed. Diseases attributed to the viroids are potato spindle tuber, exocortis of citrus, chlorotic-mottle and stunt disease of chrysanthemums.

Mycoplasma-like Organisms and Rickettsiae

Mycoplasma-like organisms have been implicated in many plant disease of the yellows type formerly considered to be of viral origin. They resemble bacteria without cell walls in form and in susceptibility to artificial culture and to antibiotic control; but they resemble viruses in their transmissibility by insects, generally leafhoppers. Corn stunt, aster yellows (Fig. 10.5), mulberry dwarf and citrus little-leaf are now believed to be caused by these organisms (Roberts and Boothroyd 1972).

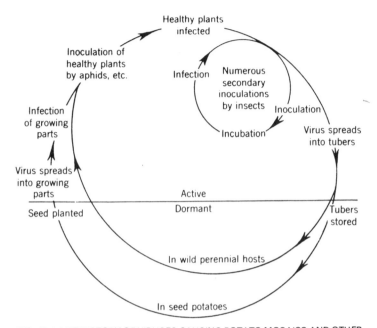

FIG. 10.4. LIFE HISTORY OF VIRUSES CAUSING POTATO MOSAICS AND OTHER
VIRUS DISEASES OF POTATOES

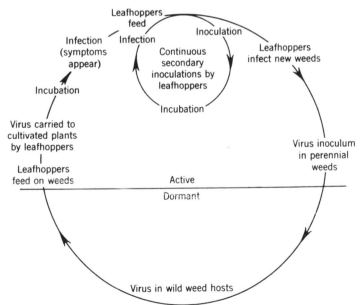

FIG. 10.5. LIFE HISTORY OF THE VIRUS (MYCOPLASMA) CAUSING ASTER
YELLOWS

Rickettsiae have also been implicated as disease-producing organisms in plants. They more closely resemble bacteria than the mycoplasma-like organisms because of their distinct cell walls, usually scalloped. Those that attack plants (e.g., Pierce's disease of grapevines) are obligate parasites inhabiting the tracheae and xylem vessels in contrast to systemic plant viruses that are translocated only in the sieve cells of the phloem.

TABLE 10.1

SOME IMPORTANT DISEASES CAUSED BY VIRUSES

Name	Method of Transmission	Host Plants
Aster yellows[1]	Leafhopper	Asters, celery, carrots, lettuce, many ornamentals and weeds
Bean mosaic	Aphids (several), seed, pollen	Beans
Citrus psorosis or scaly bark	Budding and natural root grafts	Lemon, grapefruit, orange
Cucurbit mosaic	Aphids, cucumber beetles	Cucurbits, tobacco, tomato, celery, spinach, lilies, petunias, peppers, many others including weeds
Curly top	Beet leafhopper	Beets, beans, squash, spinach, alfalfa, tomato (western yellow blight), zinnia, and many others including weeds
False blossom	Leafhopper	Cranberry
Mosaic of crucifers	Cabbage and turnip aphids	Crucifers
Potato leaf roll	Aphids, infected seed tubers	Potato
Potato mosaic (mild)	Aphids, leafhoppers, infected seed tubers, contact	Potato
Pea mosaic	Infected seed, pea aphid	Peas, clover
Peach yellows and little peach	Plum leafhopper, grafting, budding	Peach, nectarine, almond, apricot, plum
Peach X-disease	Budding (insects?)	Peach, chokecherry, nectarine, sour cherry

TABLE 10.1 *(Continued)*

Name	Method of Transmission	Host Plants
Phloem necrosis of elm	Natural root grafting, leafhopper	American elm
Ring spot	Infected seed (insects?)	Dahlia, peony, petunia, to-bacco, sweet-clover
Spindle tuber[2]	Infected seed, cutting knife, contact, aphids, fleabeetles, grass-hoppers, Colorado po-tato beetle, etc.	Potato
Spotted wilt or yellow spot	Thrips	Tomato, pine-apple, nico-tiana, zinnia
Sugar cane mosaic	Infected stalk cuttings, corn aphid	Sugar cane, corn, wild and cultivat-ed grasses
Tobacco mosaic	Handling infected plants, tools, tobacco smoking and chewing, green peach aphid	Tobacco, to-mato, pepper, petunia, and others in-cluding weeds
Wheat mosaic	Contaminated soil	Wheat, rye, bar-ley, grasses

[1]See mycoplasma-like organisms.
[2]See plant viroids.

PHYSIOPATHS AS CAUSES OF PLANT DISEASES

Probably more than 50 per cent of all plant troubles are due to impro-per environmental, nutritional, or physical conditions that may be refer-red to as *physiopaths*. The ailments caused by physiopaths are called *physiogenic* or *noninfectious* diseases. Various degrees of sickness, from mild disturbances affecting the plant or crop slightly to severe ailments resulting in loss of yield or death, may be caused by physiopaths. The severity of the trouble is determined by how far from the optimum any one or more of the disturbing factors vary. In many cases it has been noted that physiogenic diseases may make plants more susceptible to other diseases. Correction of the disturbing condition or factors generally results in recovery unless the effects have progressed too far.

Determining The Cause of Physiogenic Diseases

The signs or causes of physiogenic diseases are most elusive and often little understood or completely unrecognized. Their determination re-quires a knowledge of the cultural, nutritional, and environmental conditions that favor each crop or even each crop variety; what may be best for one type of plant may harm or even kill another plant. Since rarely in nature do plants grow under fully optimum conditions, it follows that continual improvement may be obtained by finding out and reproducing as nearly as possible for each crop variety the ideal for such factors as the nutrient balance in the soil, physical character and composition of the soil, air and water balance in the soil, soil reaction (acidity or alkalinity), soil temperatures, air temperatures, air humidity, spacing of plants, cultivation, exposure and duration of exposure to sunlight, and numerous other factors that may influence the vigor and productiveness of given crops.

It is often difficult to distinguish between physiogenic diseases and those caused by living organisms and viruses, many of the symptoms being similar. Frequently the help of soil chemists and plant physiologists is necessary to track down the physiopath through chemical analysis of the plant tissues and the soil and by experimentation. At times even these methods have defied solution as to the cause of a physiogenic disease. However, the causes of many physiogenic diseases have been determined, and some of the most common ones are listed below.

1. Lack of necessary nutrients such as nitrogen, phosphorus, potassium, or magnesium salts. *General symptoms:* stunting and foliage discoloration.

2. Shortage or unavailability of minor elements such as zinc, manganese, boron, molybdenum, and copper. *General symptoms:* foliage discoloration and internal discoloration of fruits and vegetables (Fig. 10.6).

3. Excess or unbalanced condition of nutrients. *Various symptoms:* soft vegetative growth, delayed maturity, dropping of flowers, dwarfing, foliage discoloration, and "burning" of the roots.

4. Unfavorable alkalinity or acidity of the soil. May make certain nutrients unavailable. *Symptoms:* sickly yellow, dwarfed plants with poor or no root growth.

5. Unfavorable water content of soil—waterlogging. *Symptoms:* cracking of fruits and head vegetables, yellowing, wilting, defoliation, rolling of leaves, die-back, and death (Fig. 10.7).

6. Lack of soil aeration caused by raising soil grade under trees. *Symptoms:* slow death of trees with shallow root systems.

7. Unfavorable temperatures—high temperatures, frosts, freezing temperatures. *Symptoms:* foliage crinkling, fruit scorches, cankers, tipburn, "burning" of foliage, die-back (Figs. 7.3, 7.8).

8. Air pollution—toxic industrial and (or) automotive exhaust

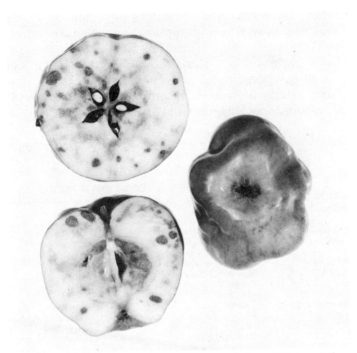

(Courtesy A. B. Burrel, Cornell University)

FIG. 10.6. INTERNAL CORK OF APPLE CAUSED BY BORON DEFICIENCY IN THE SOIL

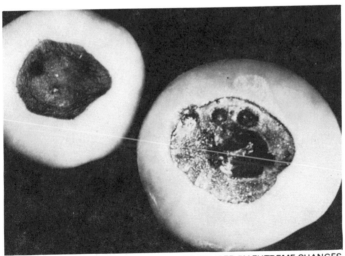

FIG. 10.7. BLOSSOM-END ROT OF TOMATO CAUSED BY EXTREME CHANGES IN WATER CONTENT OF SOIL

fumes in the air. *Symptoms:* foliage scorching, discoloration, bleaching, browning or dropping over a wide area or near industrial plants.

9. Toxic spray or dust materials. *Symptoms:* foliage discoloration, "burning," russeting of fruits, defoliation.
10. Toxic effects of certain decomposing plant residues in soil. *Symptoms:* root rots, decreased yields.

TABLE 10.2

SOME PLANT TROUBLES CAUSED BY PHYSIOPATHS

Plant Trouble	Causal Factor	Suscept
"Sand drown"	Magnesium shortage	Tobacco, cotton
"Bronzing"	Magnesium shortage	Citrus trees
Chlorosis	Magnesium shortage	Corn
Internal cork	Boron shortage	Apple
Brown heart	Boron shortage	Turnips, beets
Yellows	Boron shortage	Alfalfa
Gray speck	Manganese deficiency	Oats
Pahala blight	Manganese deficiency	Sugar cane
Chlorosis	Manganese deficiency	Ornamental trees, shrubs
Die-back	Copper deficiency	Citrus trees
"Rosette"	Zinc deficiency	Pecan, apple, pear
"Mottle leaf"	Zinc deficiency	Citrus
"Bronzing"	Zinc deficiency	Tung trees
Bud drop	Excess nitrogen plus cloudy weather	Roses, sweet peas, tomatoes
Cracking	Water excess after long dry period	Fruits and head vegetables
Blossom-end rot	Extreme changes in water content of soil	Tomatoes
Die-back	Excessive transpiration	Evergreens
Sunscald cankers	Extreme temperature drops in winter or extreme heating in summer	Deciduous trees
Leaf crinkling and puckering	Frost hitting tender leaves just unfolding	Deciduous trees and shrubs
Epinasty (downward bending of leaf stems)	Ethylene gas injury	Tomatoes
Slow death of plants	Waterlogging of soil, poor drainage, raising soil level under trees, manufactured illuminating gas leak in soil	All dry-land trees and shrubs, shallow-rooted trees, street trees, shrubs
Leaf and twig distortion and curling	Contact with 2,4-D-type weed killers	Susceptible plants
Root rots	Recently plowed under green cover crops	Strawberries

QUESTIONS FOR DISCUSSION
1. Why have plant troubles that were finally traced to viruses long been such a mystery?
2. Explain the nature of a virus.
3. Do you think virus diseases can be controlled by cultural methods?
4. What are the typical symptoms of a virus disease in a plant?
5. Why are virus diseases so difficult to control?
6. Why are sucking insects most often responsible for transmitting virus diseases?
7. Do seeds from a plant with a virus disease have the virus in them?
8. Why are virus diseases so important in potato growing?
9. Most of our certified seed potatoes come from colder northern areas. Why?
10. Why shouldn't one smoke in the presence of young tomato plants or handle them after smoking?
11. Should cuttings be made from a plant with a virus disease?
12. Can one tell from appearance alone whether or not a plant is free from a virus disease?
13. What connection is there between weeds and some of our common virus diseases?
14. Why are physiogenic diseases so prevalent under modern agricultural practices?
15. Do you think there would be less physiogenic troubles in crops planted on virgin land?
16. What is involved in determining the cause of a physiogenic disease?
17. Can a healthy looking plant actually be suffering from a physiogenic disease?
18. Plant nutrients may be present in a soil and still be unavailable to plants. Why?
19. How does a physiogenic trouble usually show up in a crop as compared to an insect or fungous disease outbreak?
20. What type of soil would tend to develop deficiencies most often? Least often?
21. Why is the pH of the soil so important to plant growth?
22. What may cause a soil to be waterlogged, with consequent plant injury?
23. What practices may lead to an unbalanced condition of soil nutrients?
24. Winter kill hits plants mainly in late winter and early spring. Why?
25. What may be done to prevent a tree from dying when it is necessary to raise the soil grade under it?
26. What locations are most subject to late spring frosts?

27. Why are newly transplanted trees susceptible to sunscald cankers?
28. What does it mean when young tomato seedlings in a home develop epinasty?
29. It may be harmful to plant a crop on land where a cover crop was just plowed under. Why?

Chapter 11

The Techniques of
Pest Control

Many diverse techniques are available for the control or suppression of pest populations harmful to humans, animals, and crops. Each technique may be used alone or integrated with others in the suppression or destruction of target pest populations (pests to be controlled) in pest management programs whose purpose is to produce more ecologically safe, efficient and economical pest control. The techniques are based on nine principles of pest control.

Nine Principles of Pest Control

Exclusion.—The prevention of the distribution and dispersal of a plant or animal pest or pathogen into an uninfested area or the prevention of its establishment there.

Eradication.—The elimination of a plant or animal pest or pathogen from an area where it has become established.

Protection.—The prevention of a pest or pathogen from producing economic injury to a crop by placing a chemical or physical barrier between the pest or the inoculum of the pathogen and the host or suscept. For plant diseases, protective measures should generally be undertaken shortly before inoculation occurs or as soon as the incubation period begins, but in any case before infection begins.

Development of Resistance.—The manipulation of the morphology or physiology of a crop plant by selective breeding or hybridization so that the pest or pathogen cannot become established.

Therapy (Physiotherapy and Chemotherapy).—Physical or chemical curative measures involving the application of heat, cold, or chemicals to destroy pests or pathogens within plants, or to destroy them on plants whose juices have been rendered toxic by the systemic absorption of a chemical.

Avoidance.—The use or manipulation of environmental factors in crop production to take advantage of the stage in the life cycle of pests or pathogenes when it can be most easily interrupted or when it is ineffectual, scarce, or absent.

Genetic Manipulation.—The use of a pest species for its own destruction by 1) introducing undesirable or incompatible genetic traits that reduce its fitness to survive and by 2) mass sterilization techniques.

Physiological Manipulation.—The use of natural and synthetic growth regulating hormones and pheromones (hormonal signals) to suppress pest populations.

Use of Biological Agents.—The reduction or suppression of pest populations by means of living organisms encouraged or manipulated by man.

PESTICIDES

Up to the present time the most common and expedient means of controlling pests involved the application of chemical compounds that kill the target pests by their chemical action. These compounds are termed pesticides when control of pests in general are considered, but more specifically they are designated as:

1. Insecticides when used for insect control.
2. Acaricides or miticides when used for mite, spider, or tick control.
3. Molluscicides when used for slug and snail control.
4. Fungicides when used for the control of disease-producing fungi.
5. Bactericides when used to kill bacteria.
6. Nematicides when used to destroy nematode worms (nemas).
7. Herbicides when used to destroy weeds or their seed.
8. Rodenticides when used for rodent control.

The role of pesticides may be either protective, the prevention of harmful organisms from producing economic injury to plant or animal crops; or eradicative, the destruction of harmful organisms after infestation or infection has occurred.

Classification of Pesticides

Pesticides may be classified according to their chief mode of action on the target pests even though they may act in more than one way.

Stomach Pesticides.—Those that are toxic to organisms when ingested, usually with their food, and take effect through the alimentary tract. Their use is limited to surface-feeding pests with chewing mouthparts. The pests need not be present when they are applied.

Contact Pesticides.—These kill by direct contact at the time of application with some external portion of the organism. They may be used for both chewing and sucking pests which must be present and contacted to be killed unless there is a residual contact effect.

Residual Contact Pesticides.—These are contact pesticides with extended residual toxicity through foot (tarsal) or body contact. Direct body contact with the pests at the time of application is not essential for control.

Fumigants.—These are pesticides that have high enough vapor pressures, natural or heat induced, to produce lethal concentrations of vapors or gases which enter primarily through the respiratory system of the organisms. They are used in enclosed spaces or the soil to destroy the pests present.

Systemic Pesticides.—These may be either plant systemics or animal systemics. Plant systemics may be insecticides, acaricides, nematocides or fungicides that are soluble enough to be absorbed harmlessly by plants through their seed, roots, stems, trunks or foliage and can be translocated by the sap to the points of attack to destroy the plant-feeding organisms. Systemics may actually function as stomach poisons on such sap-sucking pests as aphids, mites and scale insects and such chewing insects as leafminers and some borers. Systemic effectiveness may last from a few weeks up to a few months. Systemic fungicides function mainly as eradicants but some may give plants or crops varying lengths of protection from target pathogenes as well.

Animal systemics are pesticides that can be ingested harmlessly by animals (livestock) with their food or absorbed through their skin and translocated to points of attack to kill internal parasites such as horse bots, cattle grubs, and helminths (nematodes) and external blood-sucking parasites such as lice, ticks and biting flies.

Systemics offer a preferable method of plant protection by eradicating the pests or pathogenes at the point of attack and protecting the susceptible hosts from future attacks for varying lengths of time, while greatly reducing hazards to the environment. But because systemics are highly specific in their action on plant pests or pathogenes, resistant populations are likely to appear either by selection or by single gene mutation. Therefore, frequent and continuous use of systemics on crops for the control of target pests is not recommended. (See Table 12.2 for specific systemics.)

Repellents.—These are compounds that usually do not kill pests but are distasteful, malodorous or irritating enough to keep pests from feeding or to drive them away for some time when applied to their plant or animal hosts. They are used primarily in undiluted form as individual protectants on the skin or clothing of humans and the hair of animals to repel biting or annoying pests such as flies, mosquitoes, gnats, chiggers, and ticks. Those in use have a wide margin of safety to humans and animals but must be kept away from the eyes.

Repellents have limited practical use on plants. Hydrated lime, dusted on plant foliage, was once widely used on garden crops as a feeding deterrent for flea beetles and other foliage-chewing insects. 4-(Dimethyltriazeno)-acetanilide acts as a feeding deterrent to such surface feeders as cabbage worms, cotton leafworm, hornworms, boll weevil, and cucumber beetles but its practicality is questionable.

Repellents for use on individuals are formulated as aerosols, creams, lotions, powders and sticks. The repellents that protect humans and animals are usually lost rapidly by abrasion, evaporation and absorption through the skin. This limited residual effectiveness necessitates retreatments at intervals of a few hours or at best a few days with some exceptions.

TABLE 11.1

USEFUL REPELLENTS

Chemical and/or Trade Names	Pests Repelled
Benzyl benzoate	Chiggers on clothing, mange mites (spot treatment) on animals
Butoxyethoxy ethylacetate (Sta-Way)	Mosquitoes
Butoxypropylene glycol (Stabilene)	Flies on livestock
Deet, diethyltoluamide (Off, Delphene)	Mosquitoes, biting flies, sandflies, chiggers, fleas, ixodid ticks
Dimethyl carbate (Dimelone)	Eye gnats, mosquitos, fleas, chiggers
Dimethyl phthalate, DMP	Mosquitoes, chiggers, fleas, mites
Ethyl hexanediol (Rutgers 612)	Mosquitoes, biting flies, fleas, ixodid ticks

Attractants.—These are chemical substances that act as lures for insects. When they are secreted as hormonal substances by insects they are called *pheromones*. Pheromones may function as mating, food, ovopositing, alarm, aggregating or trail demarking stimulants. They are actually a form of communication or chemical signal generally limited to individuals of the same species which causes them to make oriented movements towards their source from some distance. They have been found and extracted from a number of insect species and many have been successfully compounded as synthetic chemical homologues of these natural attractants in the form of relatively simple linear olefinic alcohols or esters.

Mating or sex pheromones have great potency and species specificity. Fantastically minute quantities can effectively attract flying males or females from considerable distances downwind. Pheromones may attract only under certain weather conditions or only at certain times of day or night; e.g., Gyplure, gypsy moth pheromone, is not effective on cloudy days, southern armyworm moths respond to sex pheromones only in the early morning hours; and certain moths and mosquitoes respond only during their twilight active searching period.

Attractants or pheromones are very valuable when combined with traps and may serve to: (1) detect early infestations; (2) delineate areas of infestation; (3) sample or survey populations; (4) determine timing of control practices; and (5) reduce target pest populations. Specific sex pheromones combined with traps have been of great value in the detection and monitoring of insect populations before an infestation can enlarge and spread; e.g., the pheromones played a significant role in the elimination of the Mediterranean fruitfly (Medfly) in 1957 and its reinfestation in 1962 in Florida; and in the discovery of a Medfly invasion in 1975 in Los Angeles County, California. Pheromone trapping stations helped in its elimination within a year.

Pheromone-bait traps alone have not been used successfully for complete target pest elimination; however, it is possible to suppress target pest populations with them by: (1) mass-trapping (use of large numbers of bait traps in a limited area); (2) the disruption of normal communication (creating confusion) between the sexes to reduce mating efficiency; and (3) the combination of a pheromone with a chemical sex sterilant, bait or light traps, pesticides or insect pathogenes in an integrated control program. In using pheromone-baited traps, their color, size, shape, design, and placement in relation to the target species to be attracted are important considerations since each target species differs in its reactions to these factors.

NATURAL CONTROL OF PESTS

Harmful organisms may be held in check by many factors—those that occur naturally and are uncontrolled by man may be classified as natural controls; those that man initiates and regulates may be classified as applied controls. Natural controls are of primary importance in keeping the numbers of pests at low levels; but when these fail, as they frequently do, plagues arise, and growers must apply control measures of their own or else suffer great crop losses.

Natural Controls

Natural controls are Nature's own methods of keeping populations within bounds. They consist of physical factors such as adverse weather, climate, and physical environment and of biological factors such as predators, parasites, diseases, and food supply.

Weather and climate are probably the most important factors that directly or indirectly limit the abundance of pests. A knowledge of what is beneficial and of what is harmful to certain pest species often makes it

TABLE 11.2

SOME SYNTHETIC ATTRACTANTS AND PHEROMONES

Common and/or Trade Names	Species Attracted
Amlure	European chafer
Ammonium carbonate	Female houseflies
Anisylacetone	Melon fly, Queensland fruit fly
Butyl sorbate	European chafer
Codlelure (Cadlemone, Pherocon CM)	Codling moth
Cue-lure (Q-lure. Pherocon Qff)	Melon fly, Queensland fruit fly
Disparlure (Disparmone, Pherocon GM)	Gypsy moth
Eugenol-geraniol (ratio 9:1)	Japanese beetle
Gossyplure (Pherocon PBW)	Pink bollworm
Grandisol	Boll weevil
Grandlure (Grandamone, Pherocon BW)	Boll weevil
Gyplure	Gypsy moth
Heptyl butyrate	Yellowjackets
2,4-Hexadienyl butyrate	Yellowjackets
Looplure (Cabblemone, Pherocon CL)	Cabbage looper
Lysine	Mosquitoes and fruit flies
Medlure	Medfly (Mediterranean fruit fly)
Methyl cyclohexanepropionate and eugenol	Japanese beetle
Methyl eugenol	Oriental fruit fly
Methyl linolenate	Bark beetles
Muscalure (Muscamone)	Housefly
2-Phenyl ethanol	Cabbage looper
Pherocon FTLR	Fruit tree leafroller
Pherocon GBM (grapemone)	Grape berry moth
Pherocon LAW	Lesser appleworm
Pherocon OBLR (oblimone)	Oblique-banded leafroller
Pherocon OFM (orfamone)	Oriental fruit moth
Pherocon ABLR (redlamone)	Red-banded leafroller
Pherocon TBM	Tufted apple bud moth
Propylure	Pink bollworm
Racemate	Boll weevil
Rhindlure	Rhinocerus beetle
Siglure	Medfly, walnut huskfly
Trimedlure (Pherocon MFF)	Medfly, Natal fruitfly
Virelure	Tobacco budworm
Z-11	Red-banded leafroller, European corn borer, oblique-banded leafroller, smartweed borer

possible to predict their outbreaks. The effects of weather and climate on insects and plant diseases are discussed elsewhere. Years ago geographical features such as oceans, rivers, mountain ranges, and deserts were physical barriers to the spread of insects, but increased travel and speed of travel has decreased the importance of such barriers. Many pests may now successfully hitch-hike from continent to continent, and there is always danger that they will become established if conditions are favorable.

All the organisms that are naturally present in an environment and that attack and destroy insects or compete with them for food comprise the biological factors of control. These biological factors are of considerable importance, and sometimes they exert a greater influence than the climatic factors in reducing the abundance of pests. Probably every species of arthropod has parasites, predators, and diseases that attack it during some stage or stages in its existence, and these at times are so helpful to growers that other control measures are unnecessary.

Predators.—Predators are organisms that attack, kill, and eat other organisms but otherwise do not stay in contact with them. The most important predators of insects are other insects and closely related arthropods (Fig. 11.1). These consist mainly of numerous species of lady beetles and their larvae, which feed mostly on aphids and scale insects; lacewing flies and their larvae, known as aphid lions; syrphid fly larvae; nocturnal ground beetles and their larvae; tiger beetles; the praying mantis; robber flies; predaceous bugs; and spiders. Heavy aphid infestations are frequently completely destroyed by lady beetles, syrphid fly larvae, and aphid lions, when sufficient time is allowed. At times some insects will eat each other when competing for the food of a plant. Usually only one corn earworm is found in an ear of corn although many eggs are laid on the silks, because the survivor destroyed all the others that got in its way.

A majority of the land birds are insectivorous or partly so and are of great value in destroying insects. Some heavy insect infestations have been eliminated by birds alone. Also many small vertebrate animals are partly or wholly insectivorous and help in insect control. Among these are moles, shrews, skunks, bats, snakes, newts, lizards, toads, and fresh-water fish.

Parasites.—Parasites are organisms that live on or in the bodies of other organisms and feed on them, either greatly weakening or finally killing the host—usually when the parasite's development is complete. The most important parasites of insects are other insects. These insect parasites in their adult stage are generally minute to large wasplike insects and flies whose larvae feed within the eggs or bodies of many destructive insects. In some fields hardly a tomato hornworm escapes a tiny, braconid wasp. The hornworm shows no signs of the presence of these parasites until they emerge from its body and build conspicuous white cocoons over it (Fig. 11.2). The caterpillar then sickens and dies. The tiny egg parasite (Trichogramma minutum) is considered to be of great value because it attacks many different insect hosts, whereas most insect parasites limit their attacks to one insect species or closely related ones.

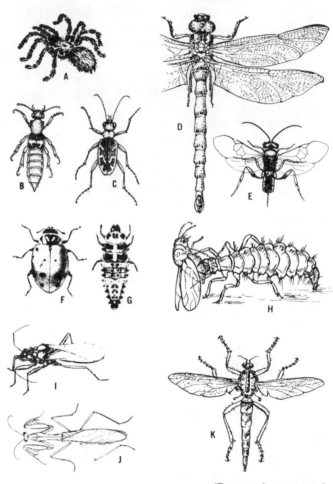

(From various sources)

FIG. 11.1. BENEFICIAL ARTHROPODS

A, Spider; B, Rove beetle; C, Tiger beetle; D, Dragon fly; E, Braconid parasite; F, G, Lady beetle and larva; H, Aphid lion; I, Assasin bug; J, Praying mantis; K, Robber fly.

Diseases.—Fungous, viral and bacterial diseases are common among insects and take heavy tolls. The fungus *Beauveria globulifera* under favorable weather conditions reduces chinch bug numbers so greatly that other control measures are unnecessary. Several species of fungi under damp hot conditions in Florida attack scale insects on citrus and

FIG. 11.2. COCOONS OF A HYMENOPTEROUS PARASITE ATTACHED TO THE
BODY OF A TOMATO HORNWORM WHICH IS SLOWLY DYING

effectively reduce their numbers. Many other fungous diseases of insects
have been noted and under the proper conditions these diseases greatly
reduce the number of a species that they attack. Bacterial and viral
diseases are also prevalent in insects, especially the larvae, and also help
in natural insect control.

APPLIED CONTROL MEASURES

Biological Control with Parasites and Predators

Natural biological control is often aided by encouraging the predators
and parasites present in a region or by bringing in and liberating new
species from other regions. For imported pests these new parasites and
predators are sought in the original home of the pests and released in the
fields to become established. The approach to this method of pest control involves: 1) the introduction of exotic species of predators or parasites; 2) the conservation of both introduced and native ones; and 3) the
augmentation of those of proven value by mass-rearing and release when
the pest is most vulnerable.

A number of pests are now partially or fully controlled by predators
and/or parasites imported from their native home. One of the first and
most striking examples of the biological control of an insect by an
imported predator was demonstrated by the elimination of the cottony-
cushion scale as a serious pest of citrus in California shortly after the
introduction of the vedalia lady beetle (*Rodolia cardinalis*) from
Australia in 1887.

The advantages to this type of control are: 1) its relative permanence
once it is established; 2) its elimination of environmental hazards; 3) its
highly economical control of the target pests—the predators and parasites do the work; and 4) its easy integration with other biological control
practices.

Biological Control with Microbial Agents

Insects are subject to many diseases caused by fungi, bacteria, and viruses. Under natural conditions these insect pathogens do not play a major role in limiting target pest populations unless environmental factors favor their multiplication and spread. Since these diseases tend to be specific for certain pests or pest families and show no indication thus far of infecting man, animals or plants, the use of these pathogens for limiting target pest populations offers an ideal method of control.

Many insect disease-producing organisms have been isolated and identified but only a relatively few so far have lent themselves to being cultured and mass-produced for use as insecticides. The microbial agents that are now produced commercially fall into two groups— bacterial agents and viral agents.

Bacterial Pathogens.—The bacterial pathogens have only two representatives both of which belong to the Bacillus or rod-shaped group of bacteria. They are: *Bacillus thuringiensis*, var. *thuringiensis, Kurstaki* or *alesti* effective for the control of many lepidopterous larvae and *Bacillus popilliae* or *lentimorbus* specific for Japanese beetle grubs in lawns and turf. Both must be ingested to take effect.

Viral Pathogens.—Many viruses pathogenic to insects have been identified—those that have DNA in their make-up are grouped as Nuclear Polyhedrosis Viruses, Granulosis Viruses, Iridoviruses, Poxviruses, Parvoviruses and Baculoviruses; and those that contain RNA are divided into Cytoplasmic Polyhedrosis Viruses, Phabdoviruses and Enteroviruses. Many others remain unclassified.

The Granulosis Viruses are the most host specific of all viruses, followed by the Nuclear Polyhedrosis Viruses. Most of the viruses have been found attacking the larvae of Lepidoptera; some infect Diptera, Hymenoptera, and Coleoptera, and a few infect other orders of insects. The only viral pesticides marketed so far belong to the Nuclear Polyhedrosis group and are registered for the control of lepidopterous larvae in cotton, tobacco and crucifers. Like the bacterial pathogens, the viral pathogens (in cystalline protein form) must be ingested by the larvae in order to produce death.

Autocide

Autocide involves the use of a pest species for its own destruction. This may be accomplished in several ways.

The Sterile Insect Release System.—This includes the mass production, sterilization by x-rays or gamma radiation and release of male insects that can compete with fertile individuals in the population. The classic example of the successful use of this technique is the virtual elimination of the screw worm (*Cochliomyia hominivorax*) in the late 1950's as a destructive pest of livestock in the southeastern states. This was accomplished by the weekly release of millions of gamma ray sterilized male flies, over an eight week period. The mating of these sterile males with the normal females in the population produced only infertile eggs with progressively increased effectiveness as the natural population declined. This approach to pest control has also been used successfully with the Oriental fruit fly (*Dacus dorsalis*) on Guam; the Mexican fruit fly (*Anastrepha ludens*) on Baja California; the melon fly (*D. Cucurbitae*) on South Pacific island of Rota; and more recently (1976) with the Mediterranean fruit fly (*Ceratitis capitata*) introduced in the Los Angeles County of California. It has also been tried with varying success in the reduction of field populations of flood-water mosquitoes in California, the codling moth in British Columbia and the boll weevil in the southeastern United States.

The method is impractical against most established pest populations of prolific species, against subeconomic species or sporadically epidemic populations. It may be practical for insect pests of considerable economic importance that lend themselves to mass-rearing and sterilization without serious effects on their vigor, longevity, behavior, or mating competitiveness. The facilities, trained personnel and cooperation at all levels of government (national and international) required, make this method of control extremely costly, if not prohibitive.

Chemosterilization.—This involves the use of chemicals that are capable of causing sexual sterility in insects when ingested or contacted by them. The sterilization may be induced by feeding them treated diets or exposing them by body or tarsal contact under mass rearing methods; or for insect species not suited for mass rearing and releasing, by exposing existing natural populations to sprays, aerosols, baits or breeding materials containing chemical sterilants.

Chemical sterilants are most effective at low population levels. The chemical sterilization of only a portion of the natural population (both sexes) theoretically tends to nullify the reproductive capacity of a like portion of the population provided competitive vigor is maintained in the sterile population. Also, theoretically the release of an overwhelming number of sterile individuals in an actively mixing pest population could come close to completely eliminating the pest by the third generation. Chemosterilants may be: 1) antimetabolites that disrupt insect nutritional and developmental process causing insects to fail to produce ova or sperm or their death, e.g., aminopterin and amethopterin; 2) alkylating agents, that prevent further development of ova or produce dominant genetic defects in sperm that prevent development of fertilized eggs, e.g., apholate, tepa, metatepa and thiotepa; and 3) the analogues of alkylating agents, e.g., hempa, hemel, thiohempa, reserpine and busulfan. Experimentally, some have been used successfully on houseflies and malarial mosquitoes. Because alkylating agents and their analogues now used are mutogenic their contact with humans and animals should be avoided. Many other compounds show promise as chemosterilants.

Genetic Manipulations.—This involves the control of a pest population by the introduction of special strains of the same pest with inheritable lethal factors or genetic weaknesses. It is the opposite of improving a plant or animal species for productiveness and involves the production or selection and mass sustained release of the pest species with undesirable genetic traits that reduce their fitness to survive.

One approach to this method of control is the manipulation of chromosomes by low doses of radiation to obtain the desirable mutations such as inability to produce viable eggs, inability to diapause, inability to fly and lack of certain essential structures in the resulting population. This method has been used successfully in some limited areas for the control of the codling moth, medfly, cabbage looper, and yellow fever mosquito.

Another approach involves the induction of sterility through inherited cytoplasmic incompatibility by the release of a selected male pest species from a different geographical region whose cytoplasm may be incompatible with the native population of the same species. This procedure has produced sterility in the F^1 hybrid malarial mosquito (*Anopheles gambiae*) and in races of the Hessian fly (*Mayetiola destructor*), and has resulted in the eradication of the mosquito *Culex pipiens fatigans* (a vector of filariasis) in 5–6 generations from a village in Burma and one near Rangoon, India.

Still another approach is to search for weakened strains of disease-producing organisms, a phenomenon called "hypovirulence," and use these strains to replace the more virulent strains of the disease. Promising results from this approach have been reported by scientists at the Connecticut Agricultural Experiment Station in the control of chestnut blight (*Endothia parasitica*). Strains of the fungus, weakened by what is suspected to be a virus, have been found which reduce the effects of the virulent fungus and allow affected trees with cankers to recover and continue growth.

Insect Growth Regulators

Insect growth regulators are hormones that control chitin synthesis, molting, growth, development, diapause, and egg maturation. The two most important are the ecdysones which initiate the post embryonic growth and development of insects and the juvenile hormones which modify events initiated by the ecdysones. Juvenile hormones prevent insects from reaching maturity by holding them in their egg, larval or nymphal stages. When they are synthesized as synthetic analogues they are referred to as *juvenoids* and when other chemicals mimic their biological activities they are called *juvenile-hormone mimics.* Juvenoids and juvenile-hormone mimics are a promising new generation of pesticides that offer the advantages of: 1) greater selectivity of action although not species specific; 2) absence of undesirable effects on humans, wild life, and the environment; and 3) compatibility with modern insect pest management principles.

Juvenoids act by contact or ingestion to alter all stages of insect growth and development generally but the last larval or nymphal stage and pupae are most sensitive and will not survive treatment.

There are some problems involved in their use:
1. They are unstable in sunlight, high water temperatures and water with high microbial populations and, therefore, must be specially formulated as solid polymers or microcapsules for slow release.
2. Applications must be timed accurately to obtain the maximum injurious effect on the most susceptible immature stage, when molting to the adult stage.
3. Not effective in mixed growth populations.
4. Evidently not insect-resistance proof, e.g., experiments in California indicated that houseflies and mosquitoes may develop resistance to juvenoids after exposure for about 15 generations.
5. Restricted in area of usefulness.

Among the promising juvenoids investigated as candidate insecticides are: the chitin synthesis inhibitors, Altosid (methoprene) and Dimilin (diflubenzuron), for such pests as floodwater mosquitoes, manure breeding flies, chironomid midges in lakes, gypsy moth larvae and boll weevil larvae. Stable-, horn- and houseflies produced sterile eggs when the adults contacted Dimilin residues applied to barn walls, and hornfly larvae in manure from cattle fed with Dimilin in drinking water failed to mature. Diflubenzuron also shows promise for stable flies, houseflies, face flies, mosquitoes and fire ants; and hydroprene (Altozar) for stored products insects. Other juvenoid candidate insecticides show some promise for sap-sucking insects on plants and biting lice on animals, e.g., tripene and kinoprene (Enstar).

Ecdysones that act as antihormones and interfere with juvenile hormone production in insects have also been found in many plants. Those designated as Precocenes I and II, isolated from the bedding plant *Ageratum haustoneanum*, have induced precocious metamorphosis, shortened the life cycle, reduced feeding and produced sterile females in the milkweed bug and cotton stainers and induced diapausing in the Colorado potato beetle.

Breeding for Pest-Resistance in Plants

The development of pest resistant plant varieties, strains, or cultivars provides the most economical, ecologically safe and labor free method of crop protection from specific pests and diseases throughout the season. A given plant variety may be immune, resistant, tolerant or susceptible to a given pest species. When a plant variety has the ability to remain completely free from attack or infestation by a specific insect or disease pathogene because of inherent genetic, structural or functional properties it is considered immune; when it has the ability to remain relatively unaffected by these organisms because of its inherent genetic and physiological or structural characteristics it is considered resistant; and when it has the ability to withstand attacks that seriously injure a susceptible host, without serious injury or crop loss, it is considered tolerant.

Plants of a given species vary in their susceptibility or resistance to harmful organisms and these characteristics may be transmitted to their progeny. This knowledge has led to the development of resistant plant varieties by:

1. Introducing a variety that through natural selection or mutation already possesses a higher than usual level of resistance to a given pest or disease.
2. Exposing plants to a given pest or disease and selecting the resistant individual plants for propagation (selective breeding).
3. Hybridizing plants to transfer the demonstrated resistant characteristics or germ plasm to desirable varieties.
4. Inducing desirable mutations for resistance.

Resistance in plants may take the form of:
1. Nonpreference in the presence of a preferred host.
2. Nonpreference in the absence of a preferred host.
3. Antibiosis—the abnormal or inhibiting effects on an organism attacking a resistant host.

There are many arthropod and disease resistant or tolerant plant varieties that have been developed through selection and breeding for crop production and many more are in the process of being developed. Growers should consult seed and nursery catalogues and their county agricultural agents or agricultural consultants on the availability of resistant seed or propagating material and make use of this built-in, season-long control of specific pests or diseases when available.

However, all is not perfect with this method of control. It has been noted that when an insect population or a disease pathogene is subjected to extreme selection pressure in the form of a resistant crop variety, those variants or mutants within the population that are able to survive on the crop may interbreed to form populations or biotypes able to overcome the resistant factors. This has already happened with a number of plant pathogenes and insect pests and has resulted in some instances in a continuous race between the plant breeders and the plant pests to keep ahead of each other, e.g., about 250 races of stem rust of wheat differing in virulence to different wheat varieties have been identified so far.

Another related problem that has cropped up is that plants bred for resistance to one pest or pathogene may be more susceptible to other pests or pathogenes, creating new problems and complicating the whole picture of plant resistance.

Physical and Mechanical Methods and Devices

These consist of methods and devices that kill or capture pests directly or exclude them without the use of pesticides. This end may be accomplished by one or more of the following methods.

Screens or Netting to Protect Plants.—These devices are practical only on a small scale where individual plants, small plant beds or valuable crops must be protected from flying insect pests (or birds) without the use of pesticides. The cloth netting placed over tobacco fields keeps out the egg-laying hornworm moths but serves primarily to improve the quality of the tobacco leaves for cigar production.

Insect Traps.—These may be light traps, color traps, bait traps or electrocuting traps (Figs. 11.3A, B and 11.4). They are most effective when used in combination with pheromones to attract, capture or kill insect pests. The essential parts of an insect trap are the attracting agent and the collecting and/or killing device.

Light traps are used for attracting and capturing photopositive night flying insect pests; however, they may capture many beneficial insects as well. Those that are equipped to emit black light (Fig. 11.3B) (near ultra and ultraviolet wave lengths) are most attractive to light sensitive insects. They are used to capture such flying insect pests as the European cornborer, corn earworm, mosquitoes, pink bollworm, cigarette beetle, European chafer, striped and spotted cucumber beetles, hornworm moths, cabbage looper and celery looper moths and houseflies in dairy barns.

Bait traps, especially pheromone-baited ones, are very successful in attracting and capturing specific night or day flying pests (see discussion and Table 11.2 under Attractants) but in general, economic control is not attainable by this device alone.

Traps that electrocute insects when they contact an electrically charged grill may employ a black light (at night) with or without a pheromone-bait; or for day flying insects a pheromone-bait as an attractant (Fig. 11.3B). When they are installed like a window screen in stables or other enclosed areas they help reduce fly populations.

Light or color may be used to repel as well as to attract insects. Light reflected from aluminum foil placed under gladiolus has been found to repel winged aphids, the vectors of mosaic, and reduce the occurrence of this virus. Yellow lights are often used in outdoor living areas to prevent attracting night flying insects, especially mosquitoes. Yellow panels on Japanese beetle traps have been found to be more attractive to the beetles than other colors. Colored panels in combination with sticky surfaces have been found effective in attracting and capturing a number of winged insect pests.

(Courtesy Detjen Corp.)

FIG. 11.3. INSECT TRAPS

A, Fly trap for use with bait; B, Combined black light and electrocuting trap.

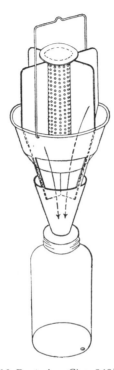

(From Rex, N.J. Dept. Agr. Circ. 242)

FIG. 11.4. A JAPANESE BEETLE TRAP, EFFECTIVE ONLY IF MANY ARE
USED OVER A WIDE AREA

Protective Barriers.—Metal *shields* properly installed on building foundations in the process of construction offer a permanent barrier to subterranean termite infestation in wooden homes and buildings.

Sticky adhesives placed in a band around tree trunks will prevent such wingless pests as spring and fall cankerworms from ascending trees to oviposit and gypsy moth larvae from ascending trees from their daytime hiding places close to or on the ground.

The use of *sticky ribbons* hung from ceilings is an old method of controlling houseflies in buildings where populations are light.

Corrugated paper bands or *burlap bands* placed around apple tree trunks will trap codling moth larvae seeking hibernating quarters. When chemically treated they will kill as well.

Protective *insect-proof packaging* will protect furs and woolens from clothes moths and carpet beetles; and dry cereals, fruits and spices from stored products pests.

Flooding and Draining.—On land where irrigation is practiced or in cranberry bogs, flooding at the right time is often useful in controlling suterranean insect pests. Flooding fields before planting in the tropics for several weeks is used to control fusarium wilt of bananas.

The draining and ditching of marshes and swamps and the elimination of all standing water not containing fish is one of the oldest and best methods of doing away with mosquitoes and some tabanid flies.

Hand Picking of Insects.—Practical only in small gardens or where labor is cheap.

Lethal Heat.—Three to four hours of exposure to dry heat at temperatures from 52°–55°C (125°–131°F) will kill all stages of insects and mites infesting granaries, mills, grain elevators, furniture, clothing, baled fibers and lumber; also nematodes and plant pathogenes in potting soils. However, damage to seed germination and furniture may result from the dessicating effect of dry heat. Small amounts of soil may be disinfested by heating in an oven to 71°C (160°F) for two hours.

Insects placed in a high-frequency electrostatic field are quickly killed by internal heat generation. This knowledge can be utilized in destroying insect infestations in grains and dry food products (provided their moisture content is low) by passing the infested material on a revolving belt through such a field. Containers that are not electrical conductors such as paper products and glassware have no shielding effects on the electrostatic waves.

Hot water dips can be used to kill disease-producing organisms within infected seed shortly before planting. The aim is to kill the pathogene without impairing the viability of the seeds; this usually requires carefully controlled conditions. The seed is placed in loose cheesecloth bags, thoroughly wetted or soaked in warm water—43.3°C (110°F), and then placed in water at the required temperature for the required time. After this treatment the seed is immediately cooled in cold water, dried, and coated with a protective dust, if necessary, before it is planted. Only fresh seed should be used; in this treatment old seed is easily injured. The following diseases may be controlled by hot water dips:

Loose smut and nematode disease of wheat—54°C (129°F, 10 min).
Brown loose smut of barley—53.3°C (128°F, 13 min).
Late blight of celery—47.7°C (118°F, 18 min).
Black leg of cabbage—50°C (122°F, 30 min).
Black leg of cauliflower—50°C (122°F, 15–20 min).
Bacterial canker of tomato—50°C (122°F, 25 min).
Root rot of aloe—46°C (115°F, 30 min).

Pressure-generated steam disinfestation of soil is a most effective method for destroying soil borne pathogenes, nematodes and arthropods. It may be applied through built-in tiles or pipes in permanent plant beds, by an inverted steam pan for confining the steam to a definite area for a given time, by covering beds with rubberized cloth or Sisalcraft paper and releasing steam underneath the cover, or in a steam-sterilizing chamber specially built for flats, pots, or soils. Efficient steam sterilization is accomplished in the time it takes to bake a medium-sized potato buried a few inches in the soil.

Cold Temperatures.—Low temperatures can be used to inhibit the development of stored products pests or to kill them outright. One to two days exposure at $-22°C$ ($-7.6°F$) will kill all insects of tropical origin and four days exposure at $-9°$ C ($+16°$ F) will kill drywood termites in furniture. Sixty day storage of apples at $0°-3°$ C ($32°-37.4°$ F) is lethal to apple maggot larvae; and 33 days exposure is lethal to plum curculio larvae.

A 90-day cold storage period for apples grown in the northeastern United States is necessary to pass quarantine regulations for shipment into California, and 12–20 days refrigeration at $1°-2°C$ ($33.8°-35.6°F$) is required at the point of shipment for fruit imported from countries having fruit fly infestation. The exact refrigeration period depends on the species of fly involved.

Summer cold storage of furs and fabrics protects them from clothes moth and carpet beetle damage, and refrigeration under cold, dry conditions is unfavorable for storage diseases of fruits and vegetables.

Humidity and Moisture.—Less than 12% moisture content of stored grain and cereal products and low relative humidity of the atmosphere in storage buildings are necessary to prevent the buildup of stored food insects, mites, and molds. Downy mildew of roses in greenhouses is controlled by increased ventilation to keep the relative humidity below 95%.

Disinfection by Aging.—In some cases seed-borne pathogens die off before the seeds lose their viability. Delaying the use of seed in these cases may then be a method of disinfecting them. Aging seed for two years destroys the fungus causing cotton anthracnose, squash foot rot, and bacterial spot of cowpea. Three-year-old celery seed is free of the inoculum causing septoria leaf spot.

Cultural Control Practices

These are the cultural practices that help the crop avoid the attack of pests. This is accomplished through cultural practices that are less favorable for the survival of the target pests in order to attain a degree of economic control. The aim is prevention of economic damage rather than the eradication of the causal organism once damage has started.

These practices are mainly useful in combination with other control practices in an integrated control program.

Sanitation.—This involves the removal, destruction, or making unfavorable the breeding and overwintering sites of plant and animal pests. This is accomplished by:

1. Burning, shredding, disking or plowing under volunteer plants, stubble and crop debris that harbor such late season or hibernating pests as the European corn borer, the pink bollworm, the boll weevil, the sugar cane borer, tobacco hornworms, and disease-producing organisms.
2. Collecting and destroying infected or infested fruit crops or pasturing livestock on remnants of crops, e.g., on corn fields for European corn borer control or on cotton fields for pink bollworm control.
3. Pruning out or surgically removing infected tree parts and coating wounds with a tree paint.
4. Removal or destruction of slash following logging operations to destroy bark beetles; and burning or debarking elm logs to prevent build-up of elm bark beetles, vectors of the Dutch elm disease fungus.
5. Keeping fields and borders free of weeds and debris that harbor plant pests and diseases or provide hibernating quarters for them.
6. Proper disposal of organic wastes such as manure, vegetable and fruit residues and garbage to limit populations of the housefly, stablefly, and Drosophila flies.
7. Good housekeeping practices to prevent infestations of stored product pests in homes, food plants, bakeries, and warehouses.

Soil Tillage.—Involves the methods and timing of soil cultivation or its avoidance to take advantage of the weak link in the life cycle of a pest:

1. Deep plowing, preferable in the spring, may bury the European corn borer and other insects hibernating in the stubble so deep that they are unable to emerge; or it may expose soil insects such as white grubs and wire worms to be eaten by birds or killed by adverse weather.
2. Fall plowing produces high mortality of the corn earworm in areas where it overwinters.
3. Fall cultivation of wheat stubble in North Dakota greatly reduces the emergence of the wheat stem sawfly.
4. Spring plowing or disking of grain stubble previously infested with grasshoppers destroys many grasshopper eggs, before they hatch, and tillage during the grasshopper hatching period to destroy all green vegetation will starve them.
5. Complete summer-fallowing of fields every second or third year in

Canada reduced wireworm populations in the soil to subeconomic levels. Allowing the summer-fallowed fields to become crusted in August and September controlled cutworms by discouraging the moths from ovipositing.

6. Complete summer-fallowing and frequent tillage of fields heavily infested with root-knot nematodes greatly reduces root knot damage to susceptible crops.

Crop Rotation.—This is the practice of following one crop with one from a different family that is not attacked by the pest (insect or pathogen) to be controlled. Crop rotation is most effective against soil pests with limited migration potential and restricted to or preferring one plant family.

1. Legumes in rotation with grass crops reduce white grub damage to the grass crop.
2. Grasses, corn or small grains followed by legumes in a crop rotation reduce damage to the legumes by the white-fringed beetle.
3. Alfalfa, crimson clover or buckwheat followed by potatoes in a crop rotation reduces wireworm damage to potatoes.
4. Not more than two successive crops of corn followed by oats or other grains will reduce damage to the corn from corn rootworms (*Diabrotica* spp).
5. Cruciferous crops (cabbage, cauliflower, broccoli, etc.) must be followed by up to six successive years of crops from other families free from mustard family weeds to help eliminate the club root disease-causing organism from the soil.
6. Pasture rotation of cattle to tick-free fields for about 65 days was a success in the eradication of the cattle tick (*Boophilus annulatus*), a one-host tick vector of bovine babeosis, in the southern United States.

Time of Planting.—Economic injury from plant pests can often be avoided by growing the crop when the pest is not present, planting it to escape the pest's greatest abundance or timing its planting so that its least attractive growth stage coincides with the egg-deposition period of the pest.

1. Early planted sweet corn in the north usually matures before migrating corn earworm and fall armyworm moths from the south arrive or become numerous. Delaying the planting of winter wheat so that it will sprout after the adult Hessian flies have emerged and died will avoid damage from this pest in the western winter wheat growing areas.
2. Late planted corn can usually escape egg deposition and serious damage from the single generation European corn borer in the

northeastern United States; but it is more severely damaged by the second brood of the two-generation strain in areas where it occurs.

3. Early planted seed of corn, beans and other vegetables on freshly manured and plowed fields may be severely damaged by the seed-corn maggot whose adults are attracted to oviposit on such fields. Late and shallow planting to encourage quick seed germination helps reduce this damage as well as damage from wireworms and seed rot fungi.

4. Delayed planting of young pines (9 months in the south, 2–3 years in the northeast) where pines have been cut or destroyed by fire avoids damage from the pales weevil Hylobius pales) and the pine root collar weevil (H. radicis).

Miscellaneous Cultural Practices

1. Selection of insect-and disease-free seed or planting stock. This may be accomplished by obtaining seed from uninfected areas, for example, bean seed from Idaho to avoid bacterial blight and anthracnose; cabbage seed from Puget Sound to avoid black rot; and certified potato seed from Maine and Canada to avoid virus diseases. Where this is not possible, seed from specially protected seed blocks or from blocks where careful inspection and roguing of diseased plants have been carried out should be utilized. Where seed or planting stock is infected, it is sometimes possible to clean out the organisms by selection, as in the case of nursery stock, bulbs, corns, or tubers, where obviously infected material may be discarded.

2. Use of certified seed or planting stock. When seed or planting stock is certified, it means that it was taken from plots that were free or nearly free of seed-borne pathogenes according to the rules of certification of each state. Generally these rules allow for only very small amounts of disease in the seed crop; or, for some diseases, none at all. The crops are inspected and rougued, if necessary, a few times during the growing season by qualified personnel. If all conditions for certification are met, the inspectors issue certification labels.

3. Avoidance of high nitrogen fertilizers on pears to reduce succulent growth which is highly susceptible to fire blight infection.

4. Destruction, replacing or planting at a distance from alternate or preferred hosts, e.g., the prevention of white pine blister rust by the eradication of Ribes spp. (currants) from the proximity of white pines; the reduction of stem rust of wheat by the eradication of common barberries; the control of cedar apple rust by the elimination of red cedars in the vicinity of apple orchards; the reduction of

grasshopper abundance in the Northern Great Plains by the replacement of broadleaved weeds along roadsides and fence rows with perennial grasses and the replacement of Russian-thistle, the chief host of the virus-bearing beet leafhopper in its breeding areas of southeast Idaho, with suitable perennial range grasses to reduce the prevalence of the curly top virus disease on susceptible crops.

5. Harvesting crops as soon as mature to reduce damage from insects and disease, e.g., sweet potatoes from the sweet potato weevil; potatoes from the tuberworm; peas from the pea weevil; cabbage from cabbage worms; and sugarcane from heavy infestations of the sugarcane borer; and tomatoes from fruit-rot diseases.

6. Avoid planting cotton or soybeans next to wheat heavily infested with the southern masked chafer to prevent feeding of the adults.

7. Harvesting before pest damage becomes extensive, e.g., early harvesting of first and second crops of alfalfa controls alfalfa weevil; harvesting the second alfalfa cutting in the early bloom or late bud stage reduces damage by the potato leafhopper; harvesting wheat before wheat stem sawfly causes severe damage and lodging; harvesting cotton before heavy boll weevil damage occurs; and harvesting timber in the fall or early winter to make it less subject to damage by borers and bark beetles.

8. Changing the pH of the soil to make it unfavorable for the development of the pathogen, e.g., increase pH of soil above 7.2 to control clubroot of crucifers; lower pH of the soil to 5.5 to control potato scab.

PEST MANAGEMENT

Definitions of Terms

An understanding of the meaning of key terminology is essential in a discussion of pest management.

Biological Control.—The action of parasites, predators, or pathogenes (naturally present in an ecosystem) on pest populations which produces a lower general equilibrium level than would prevail in the absence of these agents. When naturally occurring or new biological control agents are deliberately introduced, manipulated or modified in an agroecosystem, the term *applied biological control* is used.

Reproductive Potential.—The ability of a population of organisms to increase in density under normal conditions.

Economic Control.—The reduction or maintenance of a pest density below the economic-injury level.

Economic Damage.—The amount of crop injury that will justify the cost of applied control methods.

Economic-injury Level.—The lowest population density that will cause economic damage.

Economic Threshold.—The pest density at which control methods should be determined to prevent an increasing pest population from reaching the economic-injury level.

Ecosystem.—The interacting system comprised of all living organisms of an area and their nonliving environment. *Agroecosystem* refers to an ecosystem in an agricultural area.

Environmental Pressure.—The adverse effects on a pest population induced by both living and nonliving environmental factors.

General Equilibrium Level.—The average density of a population over a lengthy period of time in the absence of permanent environmental change.

Population.—A group of individuals of the same species that occupies a given area or ecosystem.

Selective Pesticide.—A pesticide that kills the target pests with minimal adverse effects on the ecosystem.

Many diverse procedures are available for the control or suppression of pest populations. However, most of these procedures used alone have not proven entirely satisfactory. In fact, the most widely used method of pest control, the application of compounds that kill pests by their chemical action, has created some serious problems. As a result the emphasis now is on pest management by the integration of pest control practices for a more ecologically safe, efficient and economical system of pest suppression.

Pest management may be defined as the integration of all appropriate practices, procedures and techniques relating to crop production into a single unified program or system aimed at holding one or more pests at subeconomic levels with a minimal harmful impact on the environment; or conversely the tolerance of target pest populations as long as they do not go above the economic threshold level.

Successful pest management depends on a number of important factors:

1. The correct identification of the pest and its strains, races or biotypes, if any.
2. A knowledge of the life history, activities and reproductive potential of the target pests as well as those of their natural predators and parasites.
3. An evaluation of the impact of ecological factors and the role of

natural mortality factors in the ecosystem on the reproductive potential of the pest or pests.

4. An evaluation of the different methods for a pest's control—chemical, biological, cultural and physical—with the conservation of the ecosystem and the increase of environmental pressure in mind.

5. Appropriate and practical crop inspection and population sampling techniques leading to damage predictability.

6. Determination of the economic threshold and economic injury levels for accurate timing for selective pesticide applications.

7. Determination of the most appropriate and efficient approach to the problem—integrating two or more practices, if necessary.

The concept of integrated control is actually the forerunner of the much broader concept of pest management. It was developed in the late '50s as a result of concern over the problems that developed because of our almost exclusive reliance on chemical pesticides to keep pest population in control. The concept was to integrate chemical, applied biological and natural controls to help reduce total reliance on pesticides that failed to produce long term reductions in pest populations and were creating environmental problems and health hazards. At present, it seems logical to have integrated control refer only to ways in which various techniques are used to achieve pest management. With this in mind the most appropriate definition of integrated control would be applied pest control that combines and integrates appropriate biological and chemical measures into a single unified pest control program with a minimum adverse effect on the ecosystem. This implies the judicious use of selective pesticides only when population sampling indicates that the economic threshold has been breached and economic injury levels are imminent.

Integrated control systems may also have a goal that is most acceptable to growers at this time—the reduction of pest populations to as near zero as possible or even to complete eradication; whereas the chief aim of pest management programs is to have pest populations fluctuate at levels low enough to be acceptable or tolerated. According to E.F. Knipling, formerly chief of the Entomology Research Division of the U.S.D.A., complete eradication involves "a massive coordinated attack on a pest species using a selective combination of control techniques, applied simultaneously or in a sequence with the goal of completely eradicating the population from a given major geographic region" (Barbosa and Peters 1972). The feasibility of this approach for certain major pests has been demonstrated by the use of the sterile insect release system and its success with the screw-worm problem in the Southeast and the invasions of four tropical fruit flies in citrus growing areas of the United States. However, its implementation is beyond the reach of individual growers.

"The system of insect control by the use of conventional insecticides (according to Knipling/Barbosa and Peters 1972) is highly efficient in terms of numbers of insects destroyed when the insect population is high, but highly inefficient when the population is low"; whereas "the sterile insect release system is inefficient as a means of insect population control when the natural density of the insect is high, but it becomes highly efficient when the natural population density is low". Therefore, "the integration of the conventional insecticide control system and the release of sterile insects provides a system of insect population control that is much more efficient than either system alone" or "provides a means of reversing the law of diminishing returns in dealing with the elimination of insect populations". Other integrated control systems may complement each other in a similar manner.

Successful Pest Management Programs

In some cases only one technique may be required to give outstandingly effective control of a pest, e.g., the use of the vedalia lady beetle to control the cottony-cushion scale in California citrus groves; and the control of the Hessian fly by the observance of "fly-free" planting dates or by the use of resistant wheat varieties in the West. Generally, a number of integrated techniques based on two most important criteria— *practical pest sampling methods* and *the establishment of economic injury levels* for key target pests—are required for a successful program. Such crops as citrus, tree fruits, and vegetables require economic injury levels close to zero to make them marketable.

Because of its complex nature for most crops, pest management requires the supervision and services of professionally trained people— agricultural fieldmen, agricultural consultants, county agricultural agents, entomologists or plant pathologists—to make evaluations and recommendations. Insect pest management systems developed in one area for a crop may not work on the same crop in another area because of differences in the ecosystems and climate.

Many crops, both major and minor, may lend themselves to pest management programs for important pests and diseases. Some have already been established on such crops as cotton, corn, alfalfa, tobacco, grain sorghum, sugar cane, grains, and some vegetables in the various states where they are grown. Indiana has a computerized and customized system for alfalfa pests based on weather. Growers can obtain pertinent information through the mail, by telephone or visits to the county agent's office. Most successful pest management programs require several integrated control methods. Patience and understanding is required of growers since newly initiated programs may take more than one year to show results but eventually, when natural environmental pressure on the target pests reasserts itself, more efficient, economical and hazardless crop production should result.

QUESTIONS FOR DISCUSSION

1. Growers use pesticides with the purpose of reducing target pests to zero population levels. Is this possible or desirable?
2. What advantages has a systemic pesticide over other types of pesticides?
3. Under what conditions would an insect repellent be useful on humans? On animals?
4. Is long residual effectiveness in a pesticide necessarily an attribute? Under what conditions could it present a hazard?
5. How may pheromones be used in insect control programs?
6. Why are pheromones generally specific for a species of insect?
7. Which are affected more by weather and climate: warm-blooded organisms or cold-blooded ones?
8. Why can't we depend on natural predators, parasites and diseases present in an ecosystem to control insects?
9. Would it do any good to liberate parasites or predators in an area where they are already present?
10. Why aren't fungus, bacterial, or viral diseases used more in insect pest control?
11. Would it be possible to start a disease epidemic in an insect pest population?
12. Can the principle of exclusion be used in combatting air-borne disease-producing organisms?
13. Is it possible to eradicate a disease or insect in any given area?
14. Is there any assurance that a plant variety that is resistant to a disease or insect will remain resistant?
15. Does the use of certified seed ensure a disease-free crop?
16. In what ways does crop rotation help control plant diseases? Insect pests?
17. How should one proceed to try to develop a plant variety resistant to a certain disease or insect pest?
18. Lesion nematodes do not thrive on tomatoes in soil where marigolds have been grown. Can you offer any explanation for this?
19. Why do fumigants tend to make more effective soil disinfestants than chemical solutions or suspensions?
20. How may insect traps be used in pest control programs?

21. How can heat and cold be used in insect and disease control?
22. What cultural practices can a grower apply most easily?
23. What is the mechanism by which insects or plant pathogens may overcome resistance in plants?
24. Why are some of the new spectacular approaches to pest population suppression or eradication beyond the reach of individual growers?
25. Discuss the benefits of pest management and integrated control systems.
26. Is it feasible for growers to aim at the eradication of target pests?

Chapter 12

The Chemical Pesticides I

Terminology

Toxicity.—Means "how poisonous." The inherent capability of a substance to produce injury or death.

Oral Toxicity.—"How poisonous" a pesticide is to man or animals when it is swallowed.

Dermal Toxicity.—"How poisonous" a pesticide is when it is absorbed through the skin.

Inhalation Toxicity.—"How poisonous" a pesticide is when it is inhaled.

Acute Toxicity.—"How poisonous" a pesticide is after a single exposure. It is measured as *acute oral, acute dermal,* and *acute inhalation toxicity.*

Chronic Toxicity.—"How poisonous" a pesticide is to man or animals after repeated exposures. It is measured as *chronic oral toxicity, chronic dermal toxicity,* and *chronic inhalation toxicity.*

Lethal.—Means *deadly* or *fatal.*

LD$_{50}$.—Amount (lethal dose) of toxicant or active ingredient that will kill orally or dermally 50% of the tested population. The lower the LD$_{50}$ the more poisonous the pesticide. It is usually given in milligrams per kilogram (mg/kg) of body weight.

LC$_{50}$.—Acute toxicity (lethal concentration) necessary to kill 50% of the test organisms. LC$_{50}$ values are measured in milligrams (mg) per liter (dust or mist) or in ppm (parts per million) for gas or vapor.

There is no standard method of measuring chronic toxicity. It is normally stated as the daily dose necessary to produce poisoning symptoms in a definite period of time. The chronic toxicity of the organophosphate and carbamate pesticides is measured by the reduction in the cholinesterase levels in the blood. A low level means danger.

Hazard.—The chance that harm will come to beneficial organisms, including humans, from the use of a pesticide. Do not confuse this with toxicity. *Hazard* is a function of toxicity and exposure.

Residue.—Pesticide deposit remaining on the treated crops for some time after application.

Tolerance.—The maximum amount of residue which may safely remain on a harvested crop, measured in parts per million (ppm).

THE INORGANIC PESTICIDES (Table 12.2)

Inorganic pesticides were widely used in the first half of this century but they have been almost wholly replaced by new, more effective and efficient organic pesticides. The inorganics must be ingested to be effective and, therefore, act only on chewing insects as stomach poisons. They were marketed in the form of salts of such toxic elements as arsenic, fluorine, thallium, and mercury. The most popular ones in order of their initial use were the arsenicals—Paris green, lead arsenate, and calcium arsenate—and the fluoride, cryolite. Their chief disadvantages are:

1. Toxicity limited mainly to chewing-type insects.
2. Persistence in orchard soils.
3. Development of arsenical resistance in target pests.
4. Deposition of persistent toxic residues on food crops.
5. Residue tolerance restrictions.
6. Tendency to "burn" foliage.

Those pesticides containing thallium, mercury, or selenium present so great an environmental and health hazard that their use as pesticides has been banned by the Federal Environmental Protection Agency.

Silica Gels

The silica gels or aerogels are very light and finely powdered silicates that act as dessicants. They were first used to dehydrate and preserve flowers but later were found to be effective contact insecticides when used alone or in combination with pyrethrins for the control of crawling household pests. They kill insects by absorbing waxes from their chitinous cuticle leading to dehydration and death. The silica gels are nontoxic but inhalation of the dust must be avoided.

THE ORGANIC PESTICIDES (Table 12.2)

The Petroleum Oil Pesticides

The oils used as pesticides are highly refined fractions of light lubricating petroleums in which the unsaturated hydrocarbons do not exceed 8%. They were formerly classified as "dormant" oils to be used on deciduous fruit and shade trees during their dormant period; and "summer" oils to be used on trees in leaf. Safer and more effective, superior or narrow range oils are now available for use on pests of citrus in California and Florida and on deciduous fruit and shade trees in the northeast and elsewhere. Advantages in their use are:

1. Safety to humans and animals.
2. Absence of a residue tolerance on food crops.
3. Freedom from environmental contamination.
4. Absence of development of oil-resistant races of target pests.

Types of Spray Oils.—Three types of dormant spray oils are available: (1) the paraffinic New York "superior" oils used almost exclusively in the northeastern and midwestern United States; (2) the California narrow range oils in the west; and (3) the Florida oils in the southeastern United States. All have an unsulfonated residue (UR) of 92% or better which is a measure in an oil of the saturated hydrocarbons that do not react or combine with sulfuric acid. It is the unsaturated hydrocarbons that injure plants; therefore, the higher the unsulfonated residue, the lower the unsaturated hydrocarbons and the safer the oil. The New York superior oils have a viscosity designation of 60–70 sec (Saybolt at 100° F or 38° C). The less viscuous 60-sec oils are somewhat safer and more effective than the 70-sec ones.

The oils are contact insecticides and acaricides which are used at 2–3% concentration in the dormant to the pre-bloom period to destroy mites and their eggs, mealybugs, certain scale insects, leafroller eggs, pear psyllids and some other pests (with the exception of aphid eggs) that overwinter on trees and shrubs where they may be contacted. They may be applied as synergists at .5–1% concentration in combination with other pesticides such as malathion, ethion, azinophosmethyl, or carbofenothion in summer sprays.

Spray oils may be purchased as pure oils to be emulsified with water in the spray tank or as miscible oils or oil emulsions to be diluted with water. Miscible oils are oils plus an emulsifier and have the appearance of an oil. They produce a milky emulsion on dilution with water. Oil emulsions are oils plus emulsifier plus enough water to create a creamy emulsion which becomes milky on further dilution with water.

Tank-Mixed Emulsions.—To make a tank-mixed oil emulsion, these steps should be followed:

1. Run about 25 gal. of water into the spray tank.
2. Start the engine, place the ends of the spray guns in the tank, and open them.
3. Sift in the emulsifier.
4. Pour in the oil (3 gal. per 100 gal. water is the usual dosage, but it may vary from 1–4 gal. per 100 gal. (depending on the type of oil, stage of plant growth, and the pests to be controlled).
5. Run through the pump for about 2 min. and then fill the tank with water.

6. Add any other ingredients, if necessary.

7. Keep the engine running while preparing to spray the emulsion.

The chief limitation in the use of oils is their potential to cause plant injury, mainly on certain thin-barked deciduous trees. However, using the specified oils, following label directions carefully and avoiding incompatible combinations with other pesticides such as sulfurs and captan will help to eliminate plant injury.

The Botanical Insecticides

The botanical insecticides are derived from plants or synthesized as synthetic analogues. Although plant derivatives have been used for insect control since ancient times, their popularity has increased in recent years because of the emphasis on the use of environmentally "safe" pesticides.

They break down readily on exposure to sunlight and air; generally, with the exception of nicotine, have low mammalian toxicity; require no residue tolerances on plants; and with a few exceptions have produced no build-up of target insect resistance. They act chiefly as direct contact insecticides but some also have residual contact and stomach poison effectiveness as well.

The three most useful botanicals are nicotine, pyrethrum, and rotenone (derris). Nicotine has high oral and dermal toxicity to warm-blooded animals and humans in concentrated form; whereas the pyrethrins and rotenone have low oral and dermal toxicities to them. However, both pyrethrins and rotenone are highly toxic to fish and other cold-blooded organisms and should be used with caution near bodies of water.

Nicotine.—Nicotine is a toxic alkaloid, extracted from the tobacco plant, formulated for spraying as nicotine sulfate containing 40% of the alkaloid. The addition of hydrated lime to the spray solution increases its effectiveness. It may also be used as a fumigant in greenhouses for aphids and other small sap-sucking insects. The alkaloid is absorbed in a readily combustible material and burned in smoke generators to release the fumes.

Pyrethrins and Pyrethroids.—Pyrethrins are extracted from the daisy-like flower heads of two species of chrysanthemums. They are a mixture of four analogues—pyrethrins I and II and cinerin I and II. Their ability to produce a rapid "knockdown" and paralysis of insects, their low oral mammalian toxicity, their ease in application, and effectiveness in aerosol form make them desirable in the control of household, industrial, stored food and garden pests.

Because of pyrethrins' desirable characteristics, similar synthetic analogues called pyrethroids have been synthesized and tested, with the result that several are now available with greater light stability, longer residual action and as great or greater toxicity to insects. To prevent recovery of insects from the paralyzing effects of both pyrethrins and pyrethroids and to enhance their effectiveness, synergists such as piperonyl butoxide, sulfoxide, Valone or MGK–264 are used in the formulations. These synergists help prevent recovery by inhibiting the metabolizing of the insecticides in the target pests by suppressing enzymes that detoxify them.

Rotenone.—Rotenone (derris) is obtained from the roots of the tropical leguminous plants belonging to the genera *Derris, Lonchocarpus* and *Tephrosia*. It was used as a fish poison by South American Indians long before its value as an insecticide was discovered. Fish killed by rotenone are safe to eat since it is not toxic in such dosages when ingested by warm-blooded animals. It is a useful insecticide for home gardens, because it has up to a week's residual effect on insects along with safety and an exemption from a residue tolerance on plants.

The Organochlorines (Chlorinated Hydrocarbons)

This group of pesticides typically contains carbon, chlorine, and hydrogen structurally arranged on two attached phenyl rings. Their discovery, beginning with DDT in 1939, ushered in a new era in pest control that at first promised the solution of many of our serious insect pest problems, even to the point of eradicating some pests altogether. Unfortunately, their unfavorable properties magnified by their extensive and almost exclusive use for many target pests led to such serious health and environmental hazards that their use has been banned or greatly restricted by the Environmental Protection Agency (EPA) of the Federal Government and by individual states. However, they also include methoxychlor, one of the safest and still widely used insecticides, and some environmentally safe acaricides.

Until insect resistance to DDT and its related compounds appeared, first in houseflies and later in most of the other susceptible target insects, they were found to be strikingly effective as contact, residual contact, and stomach poisons. Their residual effectiveness on lice, mosquitoes, and flies has made them of outstanding value in the field of human and animal health.

The value of this class of compounds was further reduced by their highly persistent residues that accumulate in soils, waters, animals, fish, and plants; by their biomagnification in food chains; and by their indiscriminate destruction of nontarget organisms and beneficial insects such as bees, predators, and parasites. They are also suspect of being potentially carcinogenic to humans and animals.

Mode of Action.—Although they are not highly toxic to humans and animals in single doses, they are cumulative in the body fats, so that frequent exposures by ingestion, dermal contact, or inhalation may eventually produce injurious effects. These accumulative properties of the compounds in living organisms lead to their biomagnification in food chains until harmful concentrations reach their peak in predacious fish, birds, and animals, with the consequent suppression of their populations or even the threat of their extinction. Most of the organochlorines, except for readily metabolized ones such as methoxychlor, and lindane, have been replaced by less environmentally hazardous insecticides.

The Cyclodienes

This group is similar in properties and mode of action to the organochlorines but differs structurally in containing 2 double bonds in its phenyl rings and 3-dimensional structures with steroisomers.

Like the organochlorines, they are highly persistent insecticides so stable in the soil that buildings are still protected from termite invasion some 25 years after soil treatments with such cyclodienes as chlordane or dieldrin. This persistence in the soil formerly made them valuable in the control of root-feeding insects. However, the same undesirable characteristics as possessed by the organochlorines—persistence in the environment, cumulative toxicity to humans and animals, biomagnification, target insect resistance build-up, and toxicity to nontarget organisms and all aquatic life has so reduced their usefulness that they have been banned altogether or greatly restricted in usage by the Federal Environmental Protection Agency and individual state agencies.

The Organophosphates

The organophosphates are a large and growing group of versatile pesticides derived from phosphoric acid and formulated as esters containing varying combinations of oxygen, carbon, sulfur, and nitrogen attached to phosphorus. They are related to the nerve gases of World War II and range from the very hazardous TEPP and parathion to relatively safe pesticides such as malathion and diazinon. Because they are nonpersistent and have greater selective toxicity, organophosphates are widely used as replacements for the persistent organochlorines and cyclodienes. The members of this class of pesticides kill by direct contact, residual contact, stomach poison, or fumigant action. They function as insecticides, acaricides, or nematocides. Some can be absorbed by the roots of plants and act as systemic pesticides for target insects, mites, and nematodes. Others can be fed to livestock as chemotherapeutants to destroy internal and external parasites.

Mode of Action.—The organophosphates are generally much more toxic to mammals than the organochlorines and cyclodienes. They can be hazardous to health by acute oral, dermal, and inhalation toxicity and by chronic or cumulative toxicity induced by frequent symptomless exposures through contact, ingestion, or inhalation. Each successive exposure produces increasing inhibition of the cholinesterase enzyme in the blood and, when the enzyme inhibition reaches a critical level, acute illness occurs. The lowered cholinesterase level prevents the nerves and muscles from responding normally, resulting in paralysis and possibly death.

Safety Precautions.—The Environmental Protection Agency has established reentry standards for farm workers who enter fields treated with the more acutely hazardous pesticides. Waiting periods of 24–48 hr, unless protective clothing is worn, have been designated. With all other pesticides, unprotected farm workers must wait until the spray dries or the dust settles. Those organophosphate pesticides such as malthion, phosalone, and TEPP that degrade (breakdown) readily can be used to within a few days of harvest without the danger of toxic residues; others, such as diemthoate, ethion, and trichlorfon, require up to a month or more to harvest on some crops to meet residue tolerances.

The build-up of target insect resistance to this class of compounds is also prevalent. However, the advantages of their use to growers, where resistance has not appeared, outweigh their disadvantages when all precautions in their use are followed.

The Organosulfurs

This group includes sulfonates, sulfides, sulfones, and sulfites. Generally, they have two phenyl rings with sulphur as the central atom. They act mainly as direct contact and residual contact acaricides with low insect toxicity. Many are ovicidal as well as effective on the active stages of mites. Their use presents little hazard to man and animals but their value has been limited by mite resistance build-up and plant injury problems.

The Organocarbamates

The organocarbamates are carbamic acid derivatives that act as stomach poisons and, to a lesser extent, as residual contact pesticides on a fairly broad spectrum of plant pests. Some such as aldicarb and methomyl are also effective as plant systemics on insects, mites, or nematodes. Because they are readily degraded and nonpersistent in the environment, they create a minimal environmental hazard, but are highly toxic to bees. They generally have a fairly high acute oral and dermal toxicity to mammals, and a reversible, cumulative toxicity even though they are also colinesterase inhibitors. The reason for this is that the resulting carbomyl esterase is less stable than the analagous phosphoryl esterase produced by the organophosphates; therefore, subinjurious amounts are readily degraded in the body with no cumulative effects.

One of the best known organocarbamates, carbaryl (Sevin), is widely used for leaf-chewing insects and as a lawn and garden insecticide. Another, propoxur (Baygon), is widely used for roaches and other household pests. The organocarbamates are useful for insects and mites that have built-up resistance to other pesticides, but they, in turn, are creating carbamate-resistant target pests.

The Formamidines

The formamidines are a new group of pesticides that show promise in the control of target pests resistant to the phosphates and carbamates. They were considered a relatively hazardless group of compounds nontoxic to honey bees (with no use permits and re-entry restrictions required), until laboratory tests on animals indicated that they might be carcinogenic to humans. As a result, their production and use has been indefinitely suspended.

Some of the compounds are highly effective ovicides for mites and moths and also kill their immature stages; others have shown specificity for adult mites, ticks, and aphids. Their mode of action is not fully understood. The only compound that had been registered for use is chloridimeform (Galecron, Fundal) but several others designated by code numbers are undergoing investigation.

Organic Thiocyanates

This group of compounds bears molecules with SCN endings (sulfurcyanate) in their structural formulas. They include Lethane, a relatively safe contact insecticide chiefly used now as a quick knock-down agent for flying insects in combination with other insecticides; MIT (methylisothiocyanate) used as a soil fumigant for insects, fungi, nematodes and weeds; SMDC (Vapam), actually a dithiocarbamate which converts to MIT in the soil; and DMTT (a diazine) which also converts to MIT in the soil. All three soil fumigants are moderate in toxicity, but inhalation of the fumes and skin contact should be avoided.

Dinitrophenols

The dinitrophenols include compounds with insecticidal, acaricidal, ovicidal, fungicidal and herbicidal properties. Some, such as dinitrocresol, gained wide use at one time as dormant sprays on fruit trees because of their toxicity to aphid eggs, some scale insects, and apple scab in fallen leaves. However, their undesirable properties—flammable when dry, phytoxicity, toxic residue (0 residue tolerance) and staining abilities—have greatly restricted their usage. Two foliage spray compounds, binapacryl for mites and dinocap for powdery mildew, are still in use.

Organotins

This relatively new group, trialkyl tins, may function as acaricides or fungicides or both. Three compounds are now in use—cyhexatin (Plictran) and fenbutatin oxide (Vendex) for active mite stages on fruit trees and strawberries; and fentin hydroxide (Du-Ter) for fungus leaf spot diseases on pecans, peanuts, and pears. The first two are highly selective for active mite stages but are relatively nontoxic to certain predatory mites and insects, including honey bees. They are moderate in toxicity to animals, have a residue tolerance, and must not be applied within two weeks of harvest on edible plants.

Halogenated Hydrocarbons and Other Fumigants

The majority of compounds used exclusively as fumigants, other than isothiocyanates, are volatile, hydrogenated hydrocarbons with bromine, chlorine, or fluorine as the active components. Many contact and stomach poisons (e.g., nicotine, lindane, dichlorvos) that are volatile enough naturally, or by heat induction, to produce toxic vapor concentrations may also be used as fumigants. With few exceptions, fumigants are highly toxic to most living organisms by inhalation, and a few, (e.g., hydrogen cyanide and nicotine) may also be toxic to humans through prolonged dermal contact.

There are three ways of employing fumigants:

1. As soil treatments to function as insecticides, nematocides, fungicides and/or herbicides.
2. As space treatments to function as insecticides, acaracides and/or rodenticides in homes, mills, packing plants, warehouses, ships, freight cars, industrial buildings, food establishments, granaries and grain elevators.
3. As treatments for plants or plant materials to function as insecticides and/or acaricides in greenhouses, fumigating chambers or outdoors under tents or tarpaulins.

Fumigants are available in the form of: (1) volatile solids, e.g., naphthalene, paradichlorobenzene and dichlorvos in polyvinyl resin strips; (2) reactive solids, e.g., calcium cyanide and Phostoxin which are activated by air moisture to form hydrocyanic acid and phosphine gas respectively and sodium cyanide which is activated by sulfuric acid solution to form hydrocyanic gas; and (3) volatile liquids, liquified gases under pressure, liquids in degradable capsules and gel-like substances and liquids absorbed in absorbent granules, dusts, or discs—all in pressure resistant containers, e.g., methyl bromide, chloropicrin, sulfuryl fluoride, ethylene dibromide, and hydrogen cyanide.

Advantages of Using Fumigants.—There are several advantages in using fumigants.

1. Generally wide-spectrum pesticides, destroy all active stages of pests. Methyl bromide will also kill egg and pupal stages.
2. One treatment gives rapid elimination of the pests with no mess or residue problems.
3. Penetrate all spaces inaccessible to sprays, dusts, or aerosols; e.g., methyl bromide, aluminum phosphide (Phostoxin) and sulfuryl fluoride will penetrate bags of grain and flour, grain in small bins, cotton bales, and burrows of wood borers and dry wood termites; and D-D mixture and methyl bromide will penetrate nematode galls in roots.
4. Those whose vapors have a high specific gravity (much heavier than air), e.g., methyl bromide, sulfuryl fluoride and Phostoxin, give good downward penetration in dense materials, e.g., filled grain bins, flour bins, and in the soil.

Disadvantages of Using Fumigants.—Despite the aforementioned advantages of fumigants, disadvantages should also be considered.

1. Highly hazardous by inhalation to humans and animals, fumigants may paralyze and kill very quickly. Methyl bromide produces a slow, fatal, cumulative toxicity from repeated or prolonged exposures.
2. Highly trained, experienced, and licensed operators with special permits required for indoor use. Use of toxic fumigants is banned in some cities.
3. Processed foods with considerable moisture may absorb the gases; dry food does not.
4. Buildings must be sealed and occupants must leave for several hours or as long as two days. After fumigation, buildings must be thoroughly aired and checked for absence of toxic gases.
5. Protective clothing and proper gas masks with correct filter cartridges for the fumigant used are essential for indoor treatments.
6. Reinfestation may occur immediately or shortly after treatment, since fumigants have no residual effect.
7. Fumigants are ineffective at temperatures below 13°C (55°F). Best

fumigating range is between 24°–29°C (75°–85°F) for both space and soil treatments.

8. Some, such as hydrogen cyanide in high concentrations, aluminum phosphide, and ethylene oxide, are flammable and explosive. The Phostoxin formulation (aluminum phosphide plus ammonium carbamate) and ethylene oxide plus carbon dioxide (1:9 by wt) help eliminate this danger from these two gases.

9. Most fumigants are injurious to growing plants. Methyl bromide, nicotine, and dichlorvos may be used safely under controlled conditions.

10. Many fumigants may be dangerous to handle, store, transport and release when they are liquids or gases under pressure in strong heavy metal cylinders.

11. In soil fumigation, soil temperature, moisture, after treatment and waiting period are governing considerations. Soil should be in good working condition at a temperature of 24°–29°C (75°–85°F) for effective fumigation.

12. Most fumigants have distinct warning odors, but methyl bromide and sulfuryl fluoride are odorless and colorless and require a warning gas, such as chlorpicrin (tear gas) to be incorporated in their formulation.

13. Methyl bromide may produce a persistent disagreeable odor by reacting with sulfur containing materials in home furnishings, e.g., rubber, foam rubber, leather, animal hair, and iodized salt. Phosphine gas (aluminum phosphide) under high humidities will severely corrode all metals, especially copper or copper containing alloys, equipment, and electrical apparatus.

14. Fumigation is usually more expensive than other possible methods of pest control.

THE FUNGICIDES (Table 12.2)

The chemical pesticides for plant disease control function as protectants and/or eradicants. When they reduce, remove, or eliminate disease inoculum at the source they act as eradicants; when applied to form a toxic barrier between the inoculum and the host tissue to prevent infection from beginning, they act as protectants; and when absorbed through the roots or elsewhere and translocated to the infection to eliminate it and possibly protect the plant from future infection as well, they act as systemic fungicides or chemotherapeutants.

Fungicides may be used as disinfestants, disinfectants, or protectants for seeds, tubers, bulbs, corms, and rootstocks; as disinfestants or sterilants in the soil; and as protectants, eradicants, and systemics on plants. Protectant plant fungicides must be applied often enough during inoculation periods to ensure coverage of new growth.

Fungicides are generally much less hazardous to humans and animals than insecticides; they have low oral and dermal toxicities and relatively high residue tolerance levels on food crops. Most can be used to within 15 days of harvest and do not persist to contaminate the environment.

The fungicides may be divided into two groups: (1) an inorganic group made up of the sulfur compounds, copper compounds, mercury compounds and cadmium compounds; and (2) an organic group made up of several classes of compounds.

The Inorganic Fungicides

The Copper Compounds.—The copper ion is toxic to all plant cells including fungus cells; therefore, copper compounds of very low solubility must be used to avoid plant injury when attempting to control plant diseases. Because of their low margin of safety to host plants, the copper fungicides used are of the type known as insoluble, low soluble, or fixed coppers. The coppers are wide spectrum protective fungicides of moderate to low toxicity by ingestion to man and animals. However, because of their low solubility, they tend to persist in the immediate environment, so that frequent and prolonged use may lead to a build-up of phytotoxic concentrations in the soil as has occurred in some citrus groves in Florida.

Bordeaux Mixture.—The discovery of Bordeaux mixture in the latter part of the 19th century started the impetus in the development of modern chemical disease control. Its fungicidal effectiveness was discovered when a mixture of copper sulfate (blue-stone or blue vitriol) and slaked lime was sprayed on grapes in Bordeaux, France to discourage thieves. It was noted that the disease, downy mildew of grapes, did not infect these treated plants. The combination of copper sulfate and hydrated lime forms a nearly insoluble sky-blue copper suspension which adheres well on foliage after drying to give long residual disease protection.

The modern method of making Bordeaux mixture is to wash the powdered snow or super-fine grade of copper sulfate through the strainer with the agitator running as the tank is being filled. The necessary amount of hydrated lime may also be washed through the screen or puddled in a bucket and then washed through. Some of the more common formulas for making Bordeaux mixture are 8–8–100, 6–4–100 and 4–4–100 (lb copper sulfate–lb hydrated lime–gal. water). The weaker solutions are for plants that tend to be sensitive to copper injury. Care should be taken to use fresh, finely powdered hydrated or dolomitic lime only. Once formed, Bordeaux mixture should be applied at once; its effectiveness is reduced if it is allowed to stand some time before spraying. Proprietory powdered formulations of Bordeaux mixture are obtainable, but they are inferior to homemade Bordeaux because they adhere less tenaciously.

Both copper and lime in sprays may be injurious to sensitive plants. Copper injury is shown in paling of leaf margins, dwarfing, brittleness and spotting; and lime injury on lime-sensitive plants (e.g., cucurbits, young tomato plants and strawberries) shows up in brittleness, dwarfing, leaf deformation, or discoloration.

The Low-Soluble Coppers.—Because of the tendency of Bordeaux mixture to injure plant foliage, the low soluble or fixed coppers, less injurious but less tenacious, came into general use for a wide range of fungal diseases until recent years when the organic fungicides largely replaced them.

They are basic or normal copper salts that go into suspension in water or can be made into dusts. Basically, they consist of compounds such as cupric carbonate, copper oxychloride sulfate, tribasic copper sulfate, copper zinc chromates, and copper dihydrazine sulfate (an organic). They require no special precautions when being applied.

The Sulfurs.—Elemental sulfur is one of the oldest fungicides known. It also has some value as an acaricide but its chief value is as a contact protective fungicide with eradicant-fumigant properties as well, at temperatures above 21° C (70° F) against powdery mildews. Elemental sulfur is formulated as liquid lime-sulfur, wettable sulfur, flotation sulfur (a fine sulfur paste made as a by-product of gas manufacture) and sulfur dust.

Liquid lime-sulfur, an amber colored liquid, was once widely used as a fungicide on fruits but because it is disagreeable to use and harmful to foliage when temperatures are high, it has been replaced by other fungicides. It may still be used by fruit growers that wish to "burn out" apple scab under conditions of heavy infection. Wettable sulfurs and sulfur dusts are valuable only as protectants.

Sulfur is non-toxic to humans and bees but irritates the eyes and skin. It can be heated in greenhouses to produce sulfur vapor useful to control powdery mildew but burning it produces injurious sulfur dioxide. Sulfurs should never be combined with oils or applied shortly before or after oil sprays to avoid serious damage.

Cadmium Compounds.—The cadmium-containing fungicides, both organic and inorganic, e.g., cadmium chloride, cadmium succinate, and cadmium sebacate, are good residual protectants for a number of turf diseases. Their use is confined to golf greens.

Mercury Compounds.—Mercury compounds, both inorganic and organic, were once widely used as eradicants in seed treatments for the control of seed-borne diseases of grains, vegetables, cotton, peanuts, soybeans and many other crops; also for fungus diseases of turf. Their use as fungicides has now been banned by the Environmental Protection Agency on the basis of their irreversible chronic toxicity to warm-blooded animals and persistence in the environment.

Mercury fungicides may be toxic by ingestion, dermal contact and inhalation. The greatest hazard lies in mercury's chronic or cumulative effects because of its storage in the liver, kidneys and brain with consequent irreversible damage to these organs. Mercury fungicides, because of the persistence of mercury in the environment, may accumulate in bodies of water. Fish and shellfish absorb excessive amounts from the contaminated waters and endanger the health of those using them as food. Mercury has a zero residue tolerance on all edible crops.

The Organic Fungicides

Dithiocarbamates.—The dithiocarbamate fungicides, e.g., ferbam, maneb, thiram and zineb, were introduced near the end of the first half of this century, and quickly gained a reputation for efficient and effective plant disease control. Zinc, iron, sodium, and manganese salts of dithiocarbamic acid compose the majority of the formulations which are foliage, fruit, and seed protectants for a wide range of fungus diseases on all types of crops. All have a low, acute toxicity to humans and animals and create minimal environmental hazards. However, recent findings indicate that their residues may be degraded by the heat of cooking to ethylenethiourea (ETU), a carcinogen to rats at high dosage levels. On plants, most exhibit high tenacity and a good margin of safety.

Dicarboximides.—The dicarboximides, e.g., captan, captafol and folpet, form another very useful group of protective fungicides for a wide range of diseases on food crops, seeds, and seedlings. Like the dithiocarbamates, they are safe to use, create no residue problems, and are popular for field crop, lawn, and garden use.

Oxathiins.—The oxathiins are represented by two fungicides, carboxin and oxycarboxin, which function as systemics in plants. Both show a high selective toxicity for embryo-infecting cereal smut fungi, *Rhizoctonia* spp., seed and root rot fungi, and cereal rusts. Their main use is for seed treatments on cereals and cotton, but they also show promise for other uses. Both are relatively safe to use and create no environmental problems.

Benzimidazoles.—Four systemic fungicides, benomyl, thiabendazole, and ethyl and methyl thiophanate, are representative of the benzimidazoles. The thiophanates are actually not benzimidazoles but are metabolized to that group within the host plant and the fungus. All four have gained wide use as systemic eradicants and protectants for a broad spectrum of plant diseases and can be applied as foliar sprays, seed treatments, fruit and root dips, and soil applications. Benomyl (Benlate) is especially useful in controlling hard-to-kill pathogens like *Sclerotia* spp., *Botrytis* spp., *Rhizoctonia* spp., and powdery mildews. All are very safe orally and dermally to warm-blooded animals, but they may irritate mucous membranes. They are toxic to fish. Because these systemic fungicides possess very specific metabolic toxicities, great selective pressure is exerted for resistant fungal strains to appear. Therefore, to prevent resistance build-up in target pathogens, these systemics should not be used frequently on a crop over extended periods of time.

Substituted Aromatics.—The substituted aromatics are simple benzene derivatives with diverse fungicidal action. Some are useful as protectants or eradicants to control soil, seed and turf diseases, e.g., chloroneb, diazoben and pentachloronitrobenzene; and others, e.g., clorothanonil and dichloran, are effective disease protectants on many crops and ornamentals. Exotherm, the proprietory smoke-producing formulation of chlorothanonil, gives excellent control of botrytis mold and cladosporium leaf mold of greenhouse tomatoes. One of the group, Bacticin, is used as a paste on crown galls to destroy the bacterial pathogens within. None are a threat to the environment or require special protective clothing.

Dinitrophenols.—See under Organic Pesticides.

Miscellaneous Groups of Fungicides

A number of fungicidal compounds fall into groups represented by only 1 or 2 fungicides. All are relatively safe to handle and pose no residue problems or environmental hazards.

Thiazoles.—Represented by ethazole, a protectant and eradicant soil fungicide that is useful against Pythium and Phytophthora pathogens. The chemical forms either fungicidal isothiocyanates or dithiocarbamates in the soil.

Triazines.—Represented by anilazine which is a protectant used mainly on turf diseases. It cannot be used on plants after the edible parts form and is toxic to fish.

TABLE 12.1

CATEGORIES OF TOXICITY—ACCORDING TO THE FEDERAL INSECTICIDE, FUNGICIDE, AND RODENTICIDE ACT (FIFRA)

Categories	Signal Word Required on the Label	Categories of Acute Toxicity				Eye Effects	Skin Irritation
		LD_{50}		LC_{50}	Probable Oral Lethal Dose for 150-lb (68-kg) Man		
		Oral[1] mg/kg	Dermal[2] mg/kg	Inhalation[3] mg/l			
I Highly toxic	DANGER — skull and crossbones POISON	0 – 50	0 – 200	0 – 2	A few drops to a teaspoonful	Irreversible corneal opacity at 7 days	Severe irritation or damage at 72 hr
II Moderately toxic	WARNING	over 50 to 500	over 200 to 2,000	over 2 to 20	Over 1 tsp to 1 oz (8 g)	Corneal opacity reversible within 7 days of irritation	Moderate irritation at 72 hr
III Slightly toxic	CAUTION	over 500 to 5,000	over 2,000 to 20,000	20–200	Over 1 oz to 1 pt or 1 lb (½kg)	No corneal opacity, irration reversible within 7 days	Mild irritation at 72 hr
IV Relatively nontoxic	CAUTION*	over 5,000	over 20,000	over 200	Over 1 pt or 1 lb	No irritation	No irritation at 72 hr

*None required based on acute toxicity; however, nature of product and use pattern may require appropriate precautionary statements.
[1] The dose level which will kill 50% of test animals. Minimum of 14 days observation. Animals fasted for oral studies.
[2] 24-hour exposure.
[3] The air concentration which will kill 50% of test animals exposed for a period of 1 hr. to dust or mist. Minimum of 14 days observation. Vapor or gas may be expressed in ppm.

Aliphatic Nitrogens.—These are represented by the fungicide dodine which has both protectant and eradicative properties with slight systemic action. It is selective for apple scab, pear scab, and cherry leaf spot.

Organotins.—See previous discussion under Organic Pesticides.

Formaldehyde.—This material is usually purchased as a 40% solution of the gas formalin in water. It is one of the most common soil fungicides used in greenhouses and in small plots. It is effective against damping-off fungi as well as against some others, but has little effect on soil insects and on nematode worms. For a soil fungicide, the 40% solution is diluted 1–50 by volume and is applied to the prepared soil bed at the rate of about 2 qt per sq ft. A tarpaulin or wet sacks should be placed over the treated area for 24 hr, and the odor should be gone before stock or seed is planted. For treatment of small amounts of soil in pots or flats, 2½ tablespoonfuls of commercial formalin are mixed with 1 cup of water and the solution is sprinkled over and thoroughly mixed with each bushel of soil. The treated soil is then placed in the pots or flats and covered over for a day before the seeds are sown, after which the soil should be thoroughly watered. This treatment disinfests seeds, soil, and container, all in one operation.

Antibiotic Fungicides

Antibiotic fungicides are extracts of microorganisms which inhibit the growth of, or destroy bacteria or fungi in very dilute concentrations. The most important source of antibiotics for the cure of human diseases as well as plant diseases is the microorganisms that belong to the Actinomycetes, a group of very fine filamentous soil organisms that closely resemble fungi. One species, *Streptomyces griseus*, in this group is the source of two fungicidal antibiotics, streptomycin and cycloheximide.

Streptomycin is used chiefly as a bacteriacide for bacterial diseases, such as fire blight of apples and pears and bacterial leaf spots, and requires no special safety precautions, and presents no environmental hazards. However, it does have a low residue tolerance and must not be applied closer to harvest than 50 days on apples and 30 days on pears. Cycloheximide, on the other hand, is a restricted-use pesticide toxic to a wide range of organisms, including mammals but not to bacteria. It has systemic action on rusts, powdery mildews, and some turf diseases and leaf blights.

TABLE 12.2
Pesticides[1]

INSECTICIDES, ACARICIDES, NEMATOCIDES, MOLLUSCICIDES

Common and/or Trade Names	Toxicity Rating[1]	Mode of Action[2]	Target Pests
Arsenicals			
*calcium arsenate, 1907[3]	I	SP	Chewing insects on foliage and fruits
*lead arsenate, 1892	I	SP	Same
*Paris green, 1865	I	SP	Mosquito larvae; discontinued use on crops
Fluorides			
*cryolite, 1929 (Kryocide)	IV	SP	Chewing insects on foliage and citrus fruits
Silicates			
silica aerogels, 1956 (Dry-Die, Drione)	IV	C,RC	Indoor crawling pests— roaches, silver fish, ants, fleas
Petroleum Oils			
dormant oils, 1922 (superior type, 60–70 sec)	IV	C	Mites, mite eggs, scale insects, mealy bugs on ornamentals and fruits
summer oils, 1924 (superior type, 60 sec)	IV	C	Same, but at .5–1% oil concentration
Botanicals			
beer	IV	SP	Slugs as a bait
*nicotine (sulfate), 1828 (Black Leaf 40, Nico Fume)	I	C, Fum.	Aphids, thrips, mealy bugs in greenhouses or outdoors
pyrethrins, 1800	III	C	General garden, cattle, and indoor pests
pyrethroids			
allethrin, 1949	III	C	Same as above
bioallethrin (d-trans allethrin)	III	C	Same as above
permethrin	IV	RC	Lepidopterous larvae on cotton and other major crops
resmethrin, 1968	III	C	Household, greenhouse and industrial pests
tetramethrin (Neopynamin)	IV	C	Household, animal and stored product pests
rotenone, 1848	II	SP, C	General garden pests, lice and ticks on animals
Organochlorines			
chlorobenzilate, 1952 (Acaraben)	III	RC	All stages of mites on crops
dicofol, 1957 (Kelthane)	III	RC	Active stages of mites on crops
*DDT, 1939	II	SP, RC	Wide range of insects and public health pests
*lindane, 1945 (Isotox)	II	SP, RC	Tree, drywood, fabric insects, borers, chiggers

TABLE 12.2 *(Continued)*

Common and/or Trade Names	Toxicity Rating[1]	Mode of Action	Target Pests
methoxychlor, 1944 (Marlate)	IV	SP, RC	Foliage and fruit-chewing insects, livestock parasites
*pentachlorophenol, 1936 (PCP, Penta)	II	RC	General wood preservative, termites, wood rot organisms
Perthane, 1950 (ethylan)	III	SP, C	Cabbage looper, pear psylla
*toxaphene, 1946	II	SP, RC	Livestock pests, chiggers, cotton insects; grasshoppers, army worms, cutworms on field crops

Cyclodienes

*aldrin, 1948	I	SP,RC	Termites, ants
*chlordane, 1945	II	SP,RC	Same
*dieldrin, 1948	I	SP,RC	Soil insects
*endosulfan, 1956 (Thiodan)	I	SP,RC	Wide range of insects—borers, aphids, taxus weevil
*endrin, 1951	I	SP,RC	Wide range of insects
*heptachlor, 1948	II	SP,RC	Soil and turf insects
*Kepone, 1958 (chlordecone)	II	SP	Roaches (in baits)
*mirex, 1958	II	SP	Fire ants (in baits) (not in production)

Organophosphates

acephate, 1969 (Orthene)	III	Sys.	Cotton, soybean, orn. insects
Aspon, 1963	III	RC	Lawn insects, chinch bugs
*azinphosmethyl, 1953 (Guthion)	I	RC	Broad spectrum of insects on fruits, veg. and orn.
bromophos (Brofene, Nexion)	III	SP,C	Wide range of insects and mites on crops
*carbofenothion, 1955 (Trithion)	I	SP,RC	Wide range of insects and mites, on fruit, veg., nut and fiber crops
*chlorfenvinphos, 1963 (Supona)	I	RC	Soil insects, cattle pests
*chlorpyriphos, 1966 (Dursban, Lorsban)	II	SP,RC	Wide range of insects, mites on cotton, orn. and premises
coumaphos, 1956 (Co-Ral, Baymix, Resistox)	II	Sys-A	Internal, external insects on livestock; fleas
*crotoxyphos, 1963 (Ciodrin)	II	RC	External pests of livestock, face flies
crufomate, 1959 (Ruelene)	III	Sys-A	Cattlegrubs, lice, hornflies, roundworms
*demeton, 1950 (Systox)	I	C,Sys.	Leafminers, lacebugs, mites, scales on orn. and many crops
diazinon, 1956 (Spectracide)	III	SP,RC	Wide range of insects on fruits, veg., field crops, orn., turf; chiggers, spiders
dichlorofenthion, 1956 (Hexa-nema)	II	C	Soil nematodes on orn. and turf
*dichlorvos, 1960 DDVP (Vapona)	II	SP,RC,Fum.	Fleas and ticks on pets; indoor and greenhouse insects, foliar nemas, fabric pests

TABLE 12.2 *(Continued)*

Common and/or Trade Names	Toxicity Rating[1]	Mode of Action	Target Pests
dicrotophos, 1963 (Bidrin)	I	C,Sys.	Insects on field crops, cotton, corn, cattle ticks
dimethoate, 1956 (Cygon, De-Fend, Rogor)	II	RC,Sys.	Wide range of insects and mites on orn., crops, flies in barns
dioxathion, 1955 (Delnav)	II	SP,RC	Insects on livestock, mites on crops, ticks
*disulfoton, 1956 (Di-Syston)	I	Sys.	Mite, scale "crawlers" on orn. birch leafminer
*ethion, 1959 (Ethodan)	II	RC	Wide range of insects and mites on food, fiber and other crops
ethoprop, 1963 (Profos, Mocap)	II	RC	Soil insects and nemas in turf, orn., field crops, pre- or post planting applications
*famphur, 1966 (Famfos, Warbex)	I	Sys-A	Ecto- and endoparasites, in livestock
*fensulfothion, 1957 (Desanit, Terracur)	I	RC,Sys.	Soil insects and nemas after planting
*fenthion, 1957 (Baytex)	II	RC,Sys-A	Foliar nemas, insects and mites on livestock
*fonofos, 1967 (Dyfonate)	I	RC	Soil insects in corn and other crops
*formothion, 1959 (Anthio)	II	C,Sys.	Sucking insects, fruit flies and mites
malathion, 1950	III	C	Wide range of insects and mites on fruits, veg., orn. and livestock
menazon, 1961 (Sayfos)	III	C,Sys.	Aphids on many crops
*methamidophos, 1967 (Monitor)	I	C,Sys.	Resistant insects on veg., potato; European corn borer
*methyl parathion, 1949 (Penncap-M)	I	SP,RC	Wide range of insects on crops
*mevinphos, 1953 (Phosdrin)	I	C,Sys.	Insects and mites on field crops, fruit and veg.
naled, 1956 (Dibrom)	II	SP,C	Insects and mites on many crops, mosquitoes and flies
oxydemeton methyl, 1960 (Meta-Systox-R)	II	C,Sys.	Sucking insects and mites on crops and orn.
*phenamiphos, 1969 (Nemacur)	I	C,Sys.	Nemas on fiber, food and orn. crops
*phorate, 1954 (Thimet)	I	RC,Sys.	Wide range of insects and mites on crops, orn., in soil
phosalone, 1963 (Zolone)	II	SP,RC	Wide range of insects and mites on pecans and fruit
Phosmet 1966 (Imidan)	II	RC	Wide range of insects and mites on crops and orn.
prolate, 1966 (Phosmet)	II	Sys-A	Cattle grubs
ronnel, 1954 (Korlan, Trolene)	III	RC,Sys-A	Livestock pests, flies, spiders, roaches
*TEPP, 1942 (Tetron, Vapotone)	I	C	Broad spectrum of insects and mites on crops, orn.
*terbufos, 1973 (Counter)	I	RC	Soil insects on grain crops
tetrachlorvinphos, 1966 (Gardona, Rabon)	III	SP,RC	Livestock, poultry, stored products, fabric and tree pests
trichlorfon, 1952	III	Sys-A	Broad spectrum of insects:

TABLE 12.2 *(Continued)*

Common and/or Trade Names	Toxicity Rating[1]	Mode of Action	Target Pests
(Dylox, Proxol, Dipterex, Neguvon)			leaf-miners, bugs, worms, flies
Organosulfurs			
propargite 1965 (Omite, Comite)	III	RC	Mites on crops, fruits, orn.
fenson (Trifenson)	III	C	Active mites and mite eggs on crops and orn.
tetradifon, 1954 (Tedion)	IV	RC	Mites—eggs and immature stages on fruits, nuts, veg. and orn.
tetrasul (Animert)	IV	RC	Dormant mite eggs on various crops
Organocarbamates			
*aldicarb, 1965 (Temik)	I	Sys.	Insects, mites, nemas on orn. and some crops
carbaryl, 1956 (Sevin)	II	SP,RC	Wide range of insects on orn., fruits, veg., livestock and Eriophyid mites
*carbofuran, 1969 (Furadan)	I	RC,Sys.	Insects, mites, and nemas on field crops
*formetanate, 1967 (Carzol SP)	I	C	Mites and insects on apples and pears
*methiocarb, 1962 (Mesurol)	I	SP,RC	Mites, insects, snails, slugs on a number of crops and orn.
*methomyl, 1967 (Lannate, Nudrin)	I	RC,Sys.	Wide range of insects and mites on veg. and fruit
*oxamyl, 1972 (Vydate)	I	Sys.	Soil insects and nemas on nursery crops
pirimicarb, 1969 (Pirimor)	II	C,Sys.	Aphids on a wide range of crops
propoxur, 1959 (Baygon)	II	SP,RC	Indoor insects, ticks, flies, ants, roaches; lawn insects—mole crickets, sod webworms
SMDC (Vapam) See Isothiocyanates			
Formamidines			
*chlordimeform, 1969 (Galecron, Fundal)	II	SP,RC,Fum.	Wide range of insects, ticks, all stages of mites, moth larvae, and eggs on fruit, veg. (use discontinued)
Organotins			
fentin hydroxide, 1956 (Du-Ter)	II	P	Leafspot disease on potatoes, pecans, peanuts, pears
fenbutatin oxide, 1974 (Vendex)	III	RC	Active mites on fruits
cyhexatin, 1967 (Plictran)	III	RC	Active mites on fruits and other crops
Dinitrophenols			
binapacryl, 1960 (Morocide)	II	C	All mite stages, powdery mildews
*dinitrocresol, DNOC (Elgetol, Krenite, Dinitrol)	I	C,E	Dormant mites and eggs, apple scab fungi

TABLE 12.2 *(Continued)*

Common and/or Trade Names	Toxicity Rating[1]	Mode of Action	Target Pests
*dinoseb, DNBP (DN 289, Elgetol 318)	I	C,E	Same as above
dinocap, 1949 (Karathane)	III	C,E	Powdery mildews, active mite stages

Organic Isothiocyanates

Dazomet (a diazine) (Mylone)	III	Fum.	All soil pest organisms
Lethane 384, 1932	II	C	Mosquitoes, flies— "knockdown" agent
*MIT (Vorlex)	II	Fum.	All soil pest organisms
SMDC (Vapam)	II	Fum.	All soil pest organisms

Halogenated Hydrocarbons and Miscellaneous Fumigants

*aluminum phosphide, 1930 (Phosphine, Phostoxin)	I	Fum.	Grain and stored products pests
*calcium cyanide, 1923 (Cyanogas)	I	Fum.	Greenhouse pests, rodents in burrows
*chloropicrin, 1908 (Larvacide, Picfume, Tri-clor)	I	Fum.	Soil and stored products pests; also used as a warning gas with methyl bromide
*D-D mixture, 1934 (Vidden-D, Dorlone, Telone, Nemex)	I	Fum.	Soil insects, nemas and weed seeds
*dibromochloropropane, 1955 (Nemagon, Penphene, Fumazone)	I	Fum.	Nemas in established woody plants (use discontinued)
*ethylene dibromide, 1925 (Bromofume, Dowfume W-85, W-40, Pestmaster)	II	Fum.	Nemas in soil, borers in logs, stored grain pests
*ethylene oxide (Carboxide, Oxirane)	II	Fum.	Stored products pests
*hydrogen cyanide, 1886	I	Fum.	Space fumigant for indoor and stored products pests
*methyl bromide, 1932 (Dowfume MC-2, MC-33, Brozone)	I	Fum.	All soil pests, stored products and nursery stock pests
naphthalene	III	Fum.	Clothes moths, carpet beetles
paradichlorobenzene, 1915 PDB	II	Fum.	Same as above
*sulfuryl fluoride, 1960 (Vikane)	I	Fum.	Active stages of household and stored products pests, dry-wood termites

Microbial Agents

Bacillus thuringiensis, 1963 (Dipel, Biotrol, Thuricide)	IV	SP	Lepidopterous larvae on veg., fruit, in tobacco, forests
Bacillus popilliae and *lentimorbus*, Milky disease (Doom, Japidemic)	IV	SP	Japanese beetle grubs in soil

Viral Agents

cytoplasmic polyhedrosis viruses (Viron H)	IV	SP	Lepidopterous larvae in veg., tobacco, forests
granulosis viruses	IV	SP	Same as above

TABLE 12.2 *(Continued)*

Common and/or Trade Names	Toxicity Rating[1]	Mode of Action	Target Pests
Miscellaneous Compounds			
metaldehyde, 1940 (Antimilace)	III	SP	Slugs and snails
oxythioquinox, 1962 (Morestan)	III	RC,E	All mite stages, pear psylla, powdery mildew
		Fungicides	
Inorganic Coppers			
Bordeaux mixture, 1885 (Bordo)	II	P	Foliage diseases on veg., fruits, and orn.; downy mildews, peach leaf curl, fireblight
low-soluble or fixed coppers, 1932 (Kocide, Basic-cop, Microcop, Ortho Cop 53)	III	P	Same as above
Cadmium Compounds			
*cadmium chloride (Caddy, Cad-Trete)	II	P	Golf green diseases
*cadmium succinate, 1955 (Cadminate)	II	P	Golf green diseases
*cadmium sebacate (Kromad)	II	P	Turf diseases
Inorganic Sulfurs			
liquid lime-sulfur, 1852	III	C,P,E	Foliar diseases, dormant scales and mites on fruits and orn.
sulfur dust, 1880	IV	RC,P,E	Powdery mildews, mites on fruits, orn.
wettable sulfur	IV	RC,P,E	Same as above
Dithiocarbamates			
ferbam, 1931 (Fermate, Carbamate)	IV	P	Foliar and fruit diseases, rusts, black rot of grape
maneb, 1950 (Manzate, Dithane M-22)	IV	P	Foliar diseases, black spot of roses
mancozeb, maneb-zinc ion, 1961 (Dithane M-45, Fore, Manzate 200, Dikar)	IV	P	Foliar, fruit, turf, nut, seed and seedling diseases; smuts, rusts
nabam, 1943 (Dithane D-14)	II	P	Foliar diseases
propineb, 1960 (Antracol)	IV	P	Blights, downy mildews
thiram, 1931 (Tersan, Arasan, Thylate)	III	P,E	Seed, soil, turf and foliage diseases, apple scab, brown rot
zineb, 1943 (Parzate, Dithane Z-78)	IV	P	Foliar, fruit and veg. diseases, rusts, black rot of grape
Dicarboximides			
captafol, 1961 (Difolitan)	IV	P	Foliar diseases of fruits, veg., rice seed diseases
captan, 1949 (Orthocide, Captan Seed Treater)	IV	P,E	Wide range of diseases on fruits, veg., orn., seeds and seedlings
folpet, 1952 (Phaltan)	IV	P,E	Foliar diseases on orn. fruits, powdery mildews

TABLE 12.2 *(Continued)*

Common and/or Trade Names	Toxicity Rating[1]	Mode of Action	Target Pests
Oxathiins			
carboxin, 1966 (Granox Liquid, Vitavax)	IV	Sys.	Cotton seed diseases, loose and covered smuts, and rusts on cereals
oxycarboxin, 1966 (Plantvax)	III	Sys.	Rusts on greenhouse plants
Benzimidazoles			
benomyl, 1968 (Benlate, Tersan 1991)	IV	Sys.	Wide range of diseases in seed, soil, turf, veg., fruit, orn.
ethyl thiophanate (Topsin)	IV	Sys.	Same as above
methyl thiophanate, 1969 (Fumgo, Spot, Kleen, Scott's Systemic)	IV	Sys.	Turf diseases
thiabendazole, 1962 (Mertect)	III	Sys.	Turf diseases, also storage rots of fruits, bulbs, corms, sweet potatoes
Thiazoles			
ethazole, 1969 (Terrazole, Koban, Turban)	III	P,E	Soil fungus diseases— *Pythium* spp, *Phytophthora* spp
Triazines			
anilazine, 1955 (Dyrene, Scott's No. III)	III	P	Turf diseases, leaf spots and blights on potato, tomato and curcubits
Substituted Aromatics			
chloroneb, 1965 (Demosan, Tersan SP)	IV	P,Sys	Turf, soybean and cotton seedling diseases
chlorothalonil, 1964 (Bravo, Daconil, Exotherm, Termil)	IV	P	Foliar diseases of veg., orn., turf and peanuts; Botrytis in greenhouse
dichloran, 1959 (Botran, Botec)	IV	P,E	Diseases on many crops and orn., peanut seed diseases
*pentachlorophenol see Organochlorines			
pentachloronitrobenzene, 1930's PCNB (Terrachlor)	III	P	Clubroot of crucifers in soil, damping-off of seeds (with captan)
Aliphatic Nitrogens			
dodine, 1956 (Cyprex)	III	P,E	Scab on apple, pear, pecan; cherry leaf spot; blights on sycamores, black walnuts
Dinitrophenols			
See under Pesticides, Dinitrophenols			
Fungicidal Fumigants			
formaldehyde (Formol, Formalin)	III	E	Damping-off diseases in soil; in dips for seed, tuber and bulb diseases

TABLE 12.2 *(Continued)*

Common and/or Trade Names	Toxicity Rating[1]	Mode of Action	Target Pests
Others, see under Insecticides, Organic Thiocyanates; Halogenated Hydrocarbons.			

Organotins
See under Pesticides, Organotins

Antibiotics

*cycloheximide, 1946 (Actidione, TGF, PM, or RZ)		E	Turf diseases, rusts, powdery mildews
streptomycin, 1952 (Agri-Strepp, Agrimycin, Phytomycin)		E	Bacterial diseases—fire blight, soft rot of veg.

[1]Toxicity rating based on acute oral LD$_{50}$ categories in Table 12.1.
[2]Symbols used: C, contact pesticide; E, eradicant fungicide; FUM, fumigant; ORN, ornamentals; P, protectant fungicide; RC, residual contact pesticide; SP, stomach poison; SYS, systemic plant pesticide; SYS-A, systemic animal pesticide; VEG, vegetable.
[3]Date indicates when pesticide was introduced.
*Restricted-use pesticide.

QUESTIONS FOR DISCUSSION

1. In what ways may pesticides be hazardous to humans?
2. What is the meaning of LD$_{50}$? LC$_{50}$?
3. Differentiate between toxicity and hazard in relation to pesticides.
4. Why has the use of some pesticides been banned? Restricted?
5. Is residue persistence a desirable characteristic in a pesticide?
6. Chemical disease control is sometimes likened to a man proceeding to mop up a wet floor without first stopping the water leak causing it. Explain.
7. What undesirable characteristics do the organochlorines have? The cyclodienes? The organophosphates? The organocarbamates?
8. In what ways may fumigants be used in pest control?
9. Discuss the advantages and disadvantages of using fumigants, where applicable.
10. Why is Bordeaux mixture still a useful fungicide?
11. Is a persistent residue advantageous in a fungicide?
12. Why has the use of mercury compounds been banned by the EPA?
13. Why should the use of systemic fungicides on crops be limited?
14. What are antibiotic fungicides?
15. Explain the four categories of pesticide toxicity (toxicity ratings) designated in the FIFR Act.

16. What is the difference in the cumulative (chronic) effects on humans produced by the cyclodienes as compared to those produced by the organophosphates?
17. Is fumigation practical in a large occupied apartment house?
18. Pathogens are most likely to build up resistance to systemic fungicides. Why?

Chapter 13

The Chemical Pesticides II

Pesticide Adjuvants

A pesticide adjuvant or supplement may be defined as any substance that when combined with a pesticide increases its sticking, spreading or wetting qualities; makes it safer; aids in its dilution or uniform dispersion; or increases its toxicity to the target pests. Pesticide adjuvants may be divided into (1) adhesive or sticking agents; (2) wetting and spreading agents (film extenders); (3) emulsifying agents; (4) synergists; (5) safeners or correctives and; (6) diluents or carriers.

Sticking Agents.—These substances have the function of increasing the retention of spray or dust deposits on plants by resisting the various factors involved in weathering. Proteinaceous materials such as milk products, wheat flour, calcium caseinate, blood albumin, and gelatin have the properties to act as stickers. Others are oils, gums, resins, and fine clays. When the sticker produces an elastic film with a controlled release of the pesticide in the film, it is called an *extender*. Stickers are sold under trade names such as Biofilm, Nu-film, Film Fast, Triton B1956 and Exhalt 800. Some that also function as spreaders and wetters are sold under various trade names and usually consist of sulfated alcohols, sulfates of fatty alcohols, esters of fatty acids, alkyl sulfonates, and petroleum sulfonates.

Spreaders and Wetters.—These are substances that lower the surface tension of a spray and therefore increase its spreading and penetrating power. A material that is a good spreader is normally a good wetter and vice versa. Wetting must always take place before good spreading can be obtained. Many of the commercial stickers mentioned in the previous paragraph may also have spreading and wetting properties and generally all three characteristics are of value in a spray material. Many commercial spray materials may already contain spreading and wetting agents and do not require any addition to the diluted spray. Excessive amounts will cause excessive run-off of the spray and a poor deposit; therefore, directions on the label should be carefully followed. Some commercial examples of spreaders and wetters are Biofilm, Citowett,

Atplus, Triton, Adsel, Nu-Trex, Hydrowet, Flo-Wet, Drysperse and Ad-Wet.

Emulsifying Agents.—Many pesticides are not soluble in water and must be dissolved as concentrates in oils or other organic solvents before they can be diluted with water for spraying. Since these organic solvents will not normally mix and stay dispersed in water, substances known as emulsifiers must be used with them to form stable, milky suspensions of the pesticide-containing droplets. Such a spray is called an emulsion. Emulsifying agents reduce the tendency of an emulsion to break up into its component parts by lowering the surface tension of the water, while coating the tiny solvent droplets to prevent their coalescing.

Most spreaders and wetters also are good emulsifiers, but some may produce more stable emulsions than others. Blood albumin, milk products, sulfonated petroleum oils, and vegetable or animal oil soaps are examples of emulsifiers. Some commercially available ones are Emulin, Dupanol, Agrimul, Sponto, Atlox, Solvaid and Emulphors.

Safeners or Correctives.—These are substances that are added to sprays to prevent loss of effectiveness or to reduce the danger of foliage injury. Most pesticides do not require them or they are already included in the formulation. Lime or zinc sulfate-lime combination is generally used with arsenicals to prevent the formation of phytotoxic arsenic acid or to neutralize it.

Synergists.—These are compounds, nonpesticidal or pesticidal, that when used in conjunction with a pesticide increase the toxicity of the mixture over the sum of the toxicities of its components. They are generally used in formulations of pyrethrins and pyrethroids to enhance their effectiveness and allow a substantial reduction in the amount of active ingredient necessary. Some synergists in use are piperonyl butoxide (Sesoxane), sulfoxide (Sulfox-Cide), Valone and MGK 264. They act to prevent the rapid metabolizing of the pesticides by inhibiting enzymes that detoxify them.

Diluents or Carriers.—A diluent or carrier may be any material used to dilute or decrease the amount of active ingredient in any spray or dust. In sprays, water or oils are the usual diluents. In concentrate sprays water and air is used as a diluent. In dusts, a number of different, finely ground diluents that occur naturally may be used. These diluents are classified in Table 13.1.

These dust diluents should flow freely, disperse readily, adhere to plant surfaces well, and be chemically inert. In addition they should be nonabrasive. Their bulk densities, which are an indication of their settling rates, vary from 11 lb per cu ft (Celite) to about 60 lb per cu ft (Florida limestone). The silicates are intermediate, varying from 30–45 lb per cu ft. Many of the diluents for dusts are ground so fine that 80–100% of the particles pass through a 325-mesh screen (0.83–3 μ particle size). Different diluents may influence the effectiveness of insecticides and fungicides quite markedly. Alkaline diluents generally should not be employed with the insecticides that are not compatible with alkalis but are best for making nicotine dusts. Pyrophyllite as a diluent for rotenone dusts has been shown to give better results than talcs or clays. Sulfur is valuable both as a diluent and as a fungicide in many dust formulations. Medium weight dusts (35–45 lb per cu ft) are preferred for ground-operated power dusters, but heavy dusts (45–60 lb per cu ft) are preferred by airplane or helicopter operators.

TABLE 13.1

DILUENTS OR CARRIERS

Aluminum Silicates	Alkaline Inorganics
Bentonite	Hydrated lime
Cherokee clay	Dolomite
Continental clay	Limestone
Feldspar	Florida limestone
Frianite	
Fuller's earth	**Neutral Minerals**
Kaolin	
Pyrophyllite	Gypsum (calcium sulfate)
	Sulfur
Magnesium Silicates	
	Botanical Flours
Talc	
Soapstone	Walnut shell flour
Attapulgite clay	Soybean flour
	Redwood bark flour
Diatomaceous Silicas	Peanut shell flour
Celite	

Pesticide Formulations

Toxic chemicals used for pest control generally cannot be marketed in their pure form but must be mixed or diluted with other materials to make them safe and easy to apply. A pesticide may be available in more than one of several formulations. The best formulation to use should be determined by a consideration of such factors as the nature of the target pests, the type of crop, where the pesticide is to be applied, the equipment available for application, the environmental and human hazards involved, and the relative costs.

Dusts (D).—These are finely ground pesticidal chemicals mixed with inert powdered carriers (see Diluents or Carriers) at prescribed concentrations with small amounts of anticaking and antistatic agents added. The carrier's particles should be small enough to pass through a 60-mesh screen (60 openings to the linear inch). Pesticidal dusts are generally purchased ready to apply with dusting equipment. Although easy to apply with the proper equipment, the problems of dust drift, proper weather conditions and length of residue retention limit the use of dust formulations.

Impregnated Dusts.—These are dusts made by mixing the toxic ingredient in a volatile solvent with an absorptive carrier. The solvent evaporates leaving each dust particle coated with pesticide. Less toxic ingredient and greater effectiveness favor this type of dust.

Granular Formulations (G).—These are prepared in a manner similar to impregnated dusts but the carrier (inert clays) is composed of larger particles or pellets, usually in the 15–30 mesh range, slightly coarser than granulated sugar. This type of formulation is primarily useful in soil treatments and may be applied as surface bands in the drill or furrow alone or combined with fertilizer at planting time or later to protect plant roots or to introduce systemics into plants. The formulation has also been found useful on corn where the granules collect in the whorl and leaf axils at the point of European corn borer attack. The use of granular formulations where applicable minimizes damage to foliage, reduces residues on crops, allows applications in winds up to 20 mph, increases safety in the application of highly hazardous pesticides and reduces hazards to the environment.

Wettable Powders (WP).—These are water-insoluble pesticides in powder form that contain an inert diluent and wetting agent which allows them to be wetted and dispersed as suspensions in water for spraying. Spray tanks with agitators to keep the material in suspension are essential. *Flowable (F)* or *sprayable suspensions* are actually wettable powders that are blended with a small amount of water in a mixing mill to produce a finely ground wet paste or slurry for further dilution as suspensions in sprays. They have the disadvantage of being difficult to handle.

Water-Soluble Concentrates (SP or SL).—These are pesticidal powders or liquids that will go into solution or mix with water for dilution before spraying. They do not become milky when diluted with water.

Oil-Soluble Concentrates.—These are pesticidal compounds dissolved in such organic solvents as petroleum oils or aromatic hydrocarbons that do not mix with water. They are generally used with or without further dilution in indoor pest control, or outdoors as ultra low-volume concentrates in fog generators or aerial spray equipment to deposit the toxicants with a minimum amount of oil solvent.

Emulsifiable Concentrates.—These are solutions of a toxicant and an emulsifying agent in an organic solvent such as xylene. They generally form a milky emulsion when diluted with water for spraying purposes.

Agitation is necessary to produce the emulsion but, once made, it will not readily "break-up" (separate) into oil and water phases.

Miscible Oils, Oil Emulsions.—See Petroleum Oil Pesticides.

Poison Baits.—These are mixtures (dry, granular, or liquid) of a pesticide with food attractive to the target pest or pests. They are usually placed or applied indoors or outdoors on the soil or in containers where the pests can reach them but not on plant foliage. Stomach pesticides hazardous to the environment, toxic for foliage applications, or highly toxic to the target pests without being repellent are generally used in poison baits. Mostly they are employed for the control of cutworms, grasshoppers, crickets, ants, slugs, yellowjackets, flies, and roaches.

Encapsulated Pesticides.—These are pesticides enclosed in tiny polyvinyl capsules or beads (microencapsulation) that permit a slow release and diffusion of the pesticide. The formulation is usually used in poison baits with highly hazardous pesticides to provide longer residual toxicity and increased safety to the applicators and the environment, e.g., methyl parathion (Penncap M) and mirex (Encapsulated Mirex Bait).

Aerosols.—Aerosols are pressure or heat induced suspensions of very fine liquid or solid pesticide particles in air. The liquid particles form a fog or mist in the air, whereas the solid ones form a smoke. Aerosols may be produced as cold aerosols, liquified-gas aerosols, thermal aerosol fogs, steam aerosol fogs or thermal aerosol smokes.

Cold aerosols are produced by mechanical generators of various types, electrically or gasoline operated. In these generators the insecticidal liquid concentrate is broken up mechanically (atomized) by friction, spinning, and forced air to emerge from a nozzle as a cold fog made up of extremely fine droplets. Cold aerosols have the advantage over thermal aerosols in allowing the use of insecticides such as pyrethrins and pyrethroids which are adversely affected by the high temperatures required in thermal aerosol production. Also cold aerosols are not as visible as thermal aerosols—this may or may not be an advantage depending on the situation.

Liquified-gas aerosols are produced by dissolving the toxicant in low-boiling liquids (liquified gases) such as chlorofluorocarbons or halocarbons (Freons) and placing the solution in pressure-resistant canisters or cylinder equipped with a special valve which, when opened, ejects extremely fine mistlike spray bearing the toxicant through a tiny orifice. The propellant quickly vaporizes, leaving the active ingredients suspended in the air.

Thermal aerosol smokes are made by absorbing the toxicant in a readily combustible material and placing the combination in special containers equipped with fuses for easy ignition. The smoke forced out of punched-out holes in the container carries and distributes the toxicant in the air.

Thermal aerosol fogs are produced by blowing super-heated air past a hydraulic spray nozzle emitting an oil-based spray concentrate. This produces a dense insecticidal fog.

Steam aerosol fogs are formed by passing a combined water and oil concentrate of an insecticide through coils of tubing in a combustion chamber at high temperatures. The water turns into steam under pressure which breaks up the hot oil into minute particles when forced through a nozzle with a small orifice producing an insecticidal fog.

Most contact insecticides and acaricides can be formulated as aerosols; however, in practice the use of aerosols is mostly limited to those containing compounds of relatively low mammalian toxicity such as pyrethrins, pyrethroids and dichlorvos (Vapona). Aerosols do not disperse in the air like gases since their particles are not molecular in size, unless the toxicants have a high enough vapor pressure to act as fumigants also. Aerosols therefore tend to settle out of the air at rates determined by their particle size. Indoors, fans or a directed distribution may be necessary to distribute the aerosol evenly in a space and into nooks and crevices. Outdoors, the extent of the aerosol distribution is determined by its ejection force, particle size, volume, and the prevailing wind movement, if any. High volume fog generators are usually used outdoors, but may also be used for space treatments in large warehouse type areas.

Indoors, aerosols require the use of protective masks and clothing, and may present an additional hazard if their propellents are flammable or explosive; e.g., low-boiling hydrocarbons. The chlorofluorocarbon propellents (Freons) present a potential threat to the earth's ozone layer which shields the earth's surface from harmful ultraviolet radiation and to the earth's climate by absorbing the infrared radiation from the earth and preventing it from radiating into space, according to a report from the National Research Council which recommends restrictions and possible banning of fluorocarbon aerosol use.

Chemical Seed Treatments

Chemicals may be applied to seeds for either one or both of the following reasons: (1) to eradicate disease-producing organisms carried with, on, or in seeds; and (2) to protect seeds from soil-borne pathogenes and insects. Many of the disease-producing bacteria and the fungi that cause blights, spots, and rots are carried on or in seeds, roots, bulbs, tubers, corms, or other seed stock that is purchased by growers. It is possible to waylay these hitchhiking organisms before they are introduced into the fields, if we are aware of or suspect their presence on newly purchased seed or planting stock. This may be done by various treatments aimed at destroying the disease-producing organisms with the least injury to the seed or planting stock.

When seeds or planting stock are known to be selected from disease-free plants, there is little need for seed treatments; however, certified seed (especially seed potatoes) may require additional treatments for other diseases that certification does not cover. When uncertified seed is offered for sale, the seed growers alone are in a position to know whether their seed needs treatment, and they may make it. Often it is left up to the purchaser as a means of insurance to treat seed that may carry a seed-borne disease. Some seed-borne pathogenes such as those that cause virus diseases in potatoes and dahlias and bean anthracnose are not susceptible to seed treatment, and other means of control must be used.

Seed Disinfection.—These treatments are made to kill disease-producing organisms on or within infected seed. Disinfection is generally accomplished by dipping the seed or planting stock in hot water or in chemical solutions. The aim is to kill the pathogene without impairing the viability of the seeds; this usually requires carefully controlled conditions. (See Hot Water Dips page 220.)

Protective Seed Treatments

Many disease-producing organisms that rot seeds or the bases of seedlings are nearly always present in garden soils. Some insects also attack germinating seed. When conditions are favorable for their development, they are the chief cause of seed not sprouting or of poor stands of seedlings. Seed under glass and early spring-planted seed are especially subject to attack. Protection from these soil-borne organisms may be obtained by dipping the seeds in a chemical solution or suspension, by coating them with a dry chemical dust, or by coating them with a chemically treated film by pelleting or by the slurry method. Many seed-protecting chemicals are also used as surface disinfestants on some seeds. Certified seeds still require seed protectants; disinfected seed requires protectants if the disinfectant is not also a protectant.

In the dip method the seeds are dipped into a solution or suspension, and the residue is allowed to dry on them before planting. In the dust treatment the seeds are shaken up with the dry chemical so that each seed is protected by a thin, even film of the fungicide. The amount of dust generally varies from .1–.5% of the weight of the seed treated.

Small seed packages are best treated by slitting them open and placing in them a pinch of dust on the end of a knife. The seeds in the package should then be shaken vigorously for a few minutes so that they obtain an even dust coating. For larger amounts of seeds, a covered pint or quart mason jar or special mixing equipment can be utilized. Care should be taken not to get too heavy a coating of dust on the seeds, because this may result in delayed germination or injury. All excess dust should be screened off from the seed before it is planted.

Most flower seeds tend to be retarded in germination or harmed by some chemical dust treatments. For them, soil disinfestation alone generally gives the best results against seed rots and damping-off. Seed treatment is common for many vegetables as a means of protecting them before emergence against disease-producing organisms. Some vegetable seed such as radishes and smooth-seeded peas are very resistant to decay and soil insect damage, whereas beans, corn, and cucurbits are very subject to it. Cabbage and cauliflower when sown early in the spring are also subject to seed decay. These vegetables, along with beets, tomatoes, and peppers, usually give greatly improved stands when they are treated with seed protectants.

In pelleting, a quick-drying plastic such as methyl cellulose (Methocel) is impregnated with the seed-treating fungicide and (or) insecticide and the seed is coated with the mixture by dipping or spraying. Higher dosages of seed protectants are possible by this method, because only a certain amount of dust will normally adhere to seeds. In the slurry method, the seed is mixed with a heavy suspension of the insecticide and (or) fungicide in wettable form so that on drying each seed has a protective coat of the chemical. This method has the advantage over dust coating in that it eliminates the hazard of harmful dust in the air and allows for more even and accurate dosages; thus better results may be obtained.

There are many seed protectants on the market—some are injurious to certain varieties of seeds; others are more effective for particular diseases on certain seeds or in certain soils. One should pick the seed protectant recommended for the crop and should follow directions carefully.

Pesticide Compatibility

It is frequently advantageous to combine two or more pesticides in the same spray application. When this can be done without any adverse effect on their activity or phytotoxicity the combination is said to be compatible. Incompatibility may result from: (1) the chemicals reacting to form different compounds, e.g., alkaline fungicides or adjuvants tend to decompose synthetic organic insecticides and make them ineffective; (2) the chemicals reacting to form phytotoxic mixtures, e.g., petroleum oils and sulfurs; and (3) the inability of the formulations to mix physically, e.g., oil and water based formulations when an emulsifier is missing. It is important to know what combinations are and are not compatible. Table 13.2 provides this information for many of the pesticides.

HOW TO SELECT A PESTICIDE

The number of pesticides appearing under trade names constantly increases. Many of these products are similar and are effective against many of the same pests, according to the information given on the labels. This is very confusing to the grower who is trying to find the most effective and economical material for a certain pest or groups of pests on his crops. Pests vary greatly in their susceptibility to different chemicals in formulations. Some plants also react adversely to some ingredients in pesticides. It is of great importance to read all the information on labels carefully and to know what to look for on a label when making a selection.

The analysis or active ingredient statement on a label is of primary importance as a basis of pesticide selection, not the trade name. The pesticide companies or formulators use one or more active ingredients, usually selected from a large number of basic pesticides; they mix in some other materials as diluting agents or some that their research men have found will enhance effectiveness; and then they give the formulation a trade name to distinguish it from other companies' products that contain the same active ingredients. The result is thousands of trade-named or proprietary pesticides with comparatively few different active ingredients. In order to purchase intelligently the grower should familiarize himself with the characteristics and use of the basic pesticides and should not worry about trade names as long as well-known, reliable companies have their names or brands on the products.

Cost is another consideration in the selection of a pesticide, all other things being equal. Comparative prices may be very misleading unless one realizes that price comparisons should be computed on the basis of the cost of a unit of diluted spray or dust applied on the crops and not on the basis of the cost of a unit of concentrated material purchased in the store. For example, let us compare two pesticides under different trade names but with the same active ingredients or with different active ingredients allegedly effective against the same insect pests. The first material is priced at $2 per gal., whereas the second sells at $4 per gal. Apparently the second material is twice as expensive as the first. However, the label on the first material states that it must be diluted 1 part to 200 parts of water compared with 1 part to 800 parts of water for the second material. This makes the cost $1 for 100 gal. of the first material diluted and $.50 for 100 gal. of the second material diluted. Actually, here, the apparently more expensive pesticide costs only one-half as

TABLE 13.2

INSECTICIDE-ACARICIDE-FUNGICIDE COMPATIBILITIES

INSECTICIDES, ACARICIDES

FUNGICIDES

Legend:

0 Can be mixed.
1 When mixed with water, decomposes after standing.
2 Use wettable powder forms.
3 Mix the dry pesticide in half the water, fill remainder of tank, and add oil last.
4 Sensitive to weather conditions.
5 Use emulsifiable concentrate form.
6 Phytotoxic to *Malus* and *Pyrus* spp.
7 Combination may reduce effectiveness.
8 Mix wettable powder(s) with small amount of water to form a slurry; then add emulsifiable concentrates and slurry separately to partially filled tank. Mix and fill.
? Compatibility unknown or mixture not necessary.
X Do not mix.

Compounds (row/column axis of the compatibility matrix):

carbaryl, chlordane, chlorobenzilate, demeton, Diazinon, dimethoate, endosulfan, ethion, Guthion, imidan, Kelthane, lindane, malathion, Meta-Systox-R, methoxychlor, Morestan, oil (60-70 sec), tetradifon, benomyl, Bordeaux mixture, Botran, captan, chlorothalonil, copper (fixed), Dexon, Difolatan, dinocap (Karathane), dodine, ferbam, folpet, lime-sulfur, maneb, mancozeb, PCNB, Pipron, Polyram, Terrazole, Truban, thiram, zineb

Courtesy of N.Y. State College of Agriculture

much as the apparently less expensive pesticide. The only time a more expensive pesticide is justified on the basis of actual use is when it is more effective on insects or safer as far as injury to man, animals, or plants is concerned.

It is important when selecting pesticides to keep in mind that the susceptibility of some species of insects and plants to pesticides may vary with climate and geographical location. For that reason a material found effective and safe in one part of the country may not necessarily be effective or safe in another part.

Problems Caused by the Exclusive Use of Pesticides

The massive and indiscriminate use of pesticides, especially persistent ones, is incompatible with the health and welfare of mankind and all other animal populations. It disrupts the normal checks and balances in an ecosystem and may even lead to crop failures. We must recognize the shortcomings of pesticide usage and try to reduce them. Modern crop production cannot exist without them but they should be employed only when necessary in a pest management program that also includes cultural and biological control methods. Several problems are caused by the exclusive heavy use of pesticides.

Resistance to Pesticides.—When pesticides with a specific mode of action are repeatedly and widely used for the control of a pest or a pathogene, that pest or pathogene becomes subject to extreme selective pressure. Those variants or mutants within the population, that are able to survive, interbreed and eventually build-up populations of new races, strains, or biotypes able to resist the toxic action of the pesticides. This means that they can no longer be controlled by a given pesticide at the rates normally recommended. This phenomenon was dramatically illustrated in the late 1940's by the failure of DDT to control formerly highly susceptible populations of houseflies. The pesticides act as selecting agents to eliminate the bearers of the gene for susceptibility and favor the genotypes carrying the gene for resistance.

The physiological mechanism of resistance is usually some form of detoxification, e.g., resistant houseflies, mosquitoes and several other insects have the ability to detoxify DDT, DDD and methoxychlor by rapidly breaking them down or oxidizing them to relatively harmless compounds. Resistance to one pesticide usually results in cross-resistance to its analogues (closely related compounds) but normally not to other groups of pesticides, e.g., insects resistant to the organochlorines and/or the cyclodienes may remain susceptible to organophosphates. Residual pesticides are perfect selecting agents for building-up resistant populations because they persist for long periods at selecting levels. Resistance in a population can revert to susceptibility when use of the pesticide is discontinued but it is rapidly regained when used again.

Resistance to pesticides has been found in over 250 species of arthropods and some plant disease pathogens, e.g., 91 insect species resistant to the organochlorines, 135 to the cyclodienes, 32 to the organophosphates and 5 to the carbamates. Also pyrethrum resistance has shown up in bedbugs, body lice, and houseflies in a few countries; and rotenone resistance in the Mexican bean beetle in the northeastern United States. Tetranychid mites have shown resistance to nine different groups of acaricides.

Resistance to pesticides should be expected to develop wherever fast breeding arthropods or pathogens are exposed continuously for long periods to selecting levels of residual or systemic pesticides that produce less than 100% kill, unless the gene for resistance is lacking or deleterious to survival.

Attempts to overcome pest resistance by more frequent applications or heavier dosages only aggravate the situation and lead to the endangerment of the health and welfare of humans and other nontarget organisms. Moreover, the invariable destruction of useful predators and parasites that help keep target pest populations in check may subsequently lead to the resurgence of the target pest species or secondary outbreaks of formerly sub-economic pests not affected by the pesticide applied. "Nature abhors a vacuum"—a vacant niche in the ecosystem encourages a more resistant pest species to multiply and fill the void, e.g., mites became a serious problem in orchards and citrus groves when the intensive use of DDT or its analogues apparently destroyed their predators and parasites.

Environmental Contamination.—Highly stable pesticides, although desirable for long periods of control, create environmental problems by accumulating in soils and waters in concentrations that are harmful to both plant and animal life. Most of the organochlorines and cyclodienes and to a much lesser extent other groups have the property of persistence and accumulation in the environment. For this reason the use of many has been banned or greatly restricted by the Federal Environmental Protection Agency or similar state agencies.

The continuous applications of relatively insoluble lead arsenate in some orchards over the years have led to incredible accumulations in the soil. Accumulations of DDT in orchard soils from spraying ranged up to 100 lb/acre and in some corn fields up to 10 lb/acre. Dieldrin as a soil treatment for termites stays effective for building protection up to 25 years or more. Root crops, especially carrots, can absorb the organochlorines and cyclodienes from the soil and endanger consumers. Birds, especially nestlings, feeding on contaminated insects and earthworms, have been found paralyzed or dead.

Many of our lakes, rivers, and even coastal waters have become contaminated with harmful concentrations of organochlorines or cyclodienes. Some, like TDE, Kepone and Mirex, can accumulate in fish without killing them, but may be harmful to their reproduction and the reproduction of fish-eating birds. Dosages of DDT up to 5 lb/acre do not affect mammals but 3 lb/acre will begin to affect nestling birds being fed contaminated insects. Field applications of methoxychlor, lindane, and carbaryl have no toxic effect on birds or mammals because of their rapid metabolization and excretion.

Biomagnification.—This is the progressive build-up of a pesticide residue in the bodies of organisms in a food chain with the consequence of threatened health and/or population extinction for the animals or birds at the top of the food chains. A good illustration of biomagnification occurred when the organochlorine TDE placed in a California lake to control midges at .02 ppm accumulated to 5 ppm in algae and protozoans, to 800 ppm in the fat of plankton-feeding blackfish, to 1,600 ppm in predaceous large-mouth black bass, and to 1,600 ppm in the fish-eating western greb, the top of the chain predator. The grebs stopped producing viable eggs or healthy young because the excessive residues accumulated by the birds transferred to their eggs. Similarly, toxaphene draining into ponds from agricultural lands in the Klamath Refuges of California underwent biomagnification in the following food chain: algae, daphnia, and snails 0.2 ppm, small fish 3 ppm, large fish 8 ppm, and, at the top of the food chain, fish-eating birds (pelicans, egrets and seagulls) with a lethal concentration of 40 ppm. Likewise populations of such fish-eaters or carnivores as eagles, ospreys, peregrine falcons, penguins and lake trout have been adversely affected by reproductive failures because of the biomagnification of persistent pesticides in their ecosystems.

Humans also are at the top of many food chains and, since they can also accumulate many organochlorines and cyclodienes in their body fat and milk in amounts much greater than in their diets, they are subjected to the same hazards from the biomagnification of pesticides.

QUESTIONS FOR DISCUSSION
1. Under what conditions would a sticking agent be undesirable in a spray?
2. What would be the result of adding too much spreader-wetter to a spray?

3. Is the addition of a spreader-wetter important in concentrate spraying?
4. When would an emulsifier be necessary in spraying?
5. Why do pyrethrins or pyrethroids generally require synergists?
6. How is air used as a diluent in concentrate spraying?
7. Is a spreader-wetter most valuable in a stomach poison, contact insecticide, or a protective fungicide?
8. Under what conditions would a quick-settling dust be useful? A slow-settling dust?
9. What problems limit the use of dust formulations?
10. How are granular formulations used?
11. Which are safer as foliage sprays, wettable powders or emulsifiable concentrates?
12. What advantages have poison baits over sprays?
13. Discuss how aerosols are produced.
14. Why don't aerosols disperse like gases?
15. For what types of pest control situations are aerosols best suited?
16. Does the use of certified seed ensure a disease-free crop?
17. What types of insects and diseases can be prevented from damaging seeds by chemical seed treatments?
18. Does seed treatment alone ensure a crop free from the pathogens that are carried on or in the seed?
19. One-year-old celery seed may be heavily infested with spores of septoria leaf spot, but three-year-old seed generally does not carry the pathogens. What is the explanation?
20. Under what conditions is seed disinfection necessary? Seed protection? Both?
21. What may be some of the reasons for poor stands of seedlings?
22. Is an introduced soil-borne disease more apt to produce damage in a sterilized soil or an unsterilized soil? Explain.
23. What may happen when two incompatible pesticides are mixed in a spray solution and sprayed on plants?
24. Discuss the problems caused by the exclusive heavy use of pesticides.
25. Explain biomagnification of a pesticide in a food chain.
26. Explain the mechanism of pesticide resistance build-up in an insect population.

Chapter 14

Pesticide Legislation

REGULATORY CONTROL

Regulatory control of pests involves laws and regulations to prevent the entry and establishment of potential foreign plant and animal pests, and to eradicate, contain or suppress those that manage to gain entrance. In the early history of the United States, there were practically no legal restrictions to prevent the introduction of pests from other regions of the world. As a consequence many of our destructive pests today are of foreign origin.

The Federal Insect Pest Act of 1905 and the Plant Quarantine Act of 1912 were the first to provide for enforcement of regulations necessary to prevent the introduction of arthropod pests and plant diseases from foreign countries and to provide for the establishment of interstate quarantines to contain those that had already gained a foothold. Since then a number of additional acts and amendments have been added to the initial regulations to strengthen them.

Thousands of potential pests have been intercepted yearly at ports of entry because of these regulations. But some like the Japanese beetle, the cereal leaf beetle, the face fly, the Mediterranean fruit fly, the spotted alfalfa aphid, the imported fire ant and the Dutch elm disease pathogene have slipped through to cause severe damage in some areas, until they were contained, suppressed or eradicated.

The responsibility for regulatory control to enforce quarantine regulations in the United States lies primarily with the U.S. Department of Agriculture, in cooperation with state and local agricultural agencies. Many foreign governments also cooperate in the enforcement of plant quarantines under a 1951 agreement entitled "International Plant Protection Convention." State agencies are responsible for imposing and enforcing quarantines within their own borders and may also have regulations governing the shipment or entrance of raw commodities from other foreign or domestic sources; e.g., California has very stringent regulations governing the introduction of raw fruits, vegetables, nursery stock and propagative materials from other states and foreign countries, and confiscates all such products at all borders and ports of entry unless previously certified free of potential pests.

Quarantine regulations generally involve such requirements as:
1. Designation of quarantined areas, regions or countries with potential pests.
2. Compliance with all the requirements for certifying raw agricultural products from infested areas as pest-free at the point of origin.
3. Approved handling or treatments of commodities at points of origin to render them pest-free for certification.
4. Quarantine and inspection of agricultural commodities that may harbor pests at points of entrance with either approved disinfection or disinfestation treatments if necessary, or with confiscation and destruction.
5. The complete banning of shipments from infested regions to pest-free regions in some situations.

Copies of quarantine laws and regulations are available to shippers, exporters, and importers from agricultural agencies in the countries or states involved.

REGULATIONS FOR THE SALE, HANDLING AND USE OF PESTICIDES

Federal Pesticide Laws

The Federal Government has set standards for pesticide handling and use of pesticides. Individual states may also establish standards more rigid but not more permissive than those of the Federal government.

The Federal Insecticide, Fungicide and Rodenticide Act (FIFRA) has undergone many amendments and revisions since the original Insecticide Act of 1910 was passed to prevent the adulteration and misbranding of insecticides and fungicides. In 1947, it was revised to include herbicides and eight definite label requirements. In 1972, it was amended by the Federal Environmental Pesticide Control Act (FEPCA). It is administered by the Environmental Protection Agency (EPA) established in 1970. This law now:
1. Requires the registration of all pesticides by EPA except in certain special cases.
2. Requires the classification of all registered pesticides either as "general use" pesticides which can be used by any responsible person or as "restricted use" pesticides if the environment or user could be harmed even if directions are followed.
3. Requires that the users of restricted pesticides *must be certified* either as *"private"* or *"commercial" applicators*. Certification is to be handled by the individual states.

4. Establishes tolerances for pesticide residues that may persist on treated food crops.
5. Provides penalties for "use inconsistent with the labeling" of a pesticide.
6. Makes it illegal to store or dispose of pesticides or pesticide containers other than as directed by regulations and provides penalties for illegal use of containers.
7. Provides for civil penalties in the form of fines when the violation of a regulation is unintentional.
8. Provides criminal penalties when the law is knowingly violated in the form of fines and prison sentences.
9. Permits states to establish stricter standards but not more permissive standards.

Certification

Anyone using "restricted use" pesticides must be registered and certified by passing tests to show competence in general and in their respective fields. An applicator must be under the "direct supervision" of someone who is certified but does not necessarily have to be certified himself.

There are two classifications of certified applicators:

Private Certified Applicators.—Such persons can use or supervise the use of "restricted use" pesticides to produce agricultural commodities on property owned by himself or his employer; or can use such pesticides on the property of others on a work exchange basis. The term "agricultural commodity" covers any plant or animal product that is produced for use by man or animals. Examples: fruits, vegetables, grains, sod, nursery stock, cattle, poultry, milk, and eggs.

Commercial Certified Applicators.—These include ten different categories each of which can be divided further or omitted by states if it does not apply to their conditions. Commerical applicators may apply for certification in any or all of the categories. These ten categories are:

Agricultural Pest Control on Plants (crops) and Animals including their areas of confinement.

Forest Pest Control on forests, forest nurseries, and forest seed producing areas.

Ornamental and Turf Pest Control.

Seed Treatment (Commercial).

Aquatic Pest Control.

Right-of-Way Pest Control.

Industrial, Institutional, Structural and Health-Related Pest Control.

Public Health Pest Control. Programs carried out by State, Federal or other governmental employees.

Regulatory Pest Control. Control of regulated pests by state, federal or other governmental employees.

Demonstration and Research Pest Control involves those who use or demonstrate the use of restricted-use pesticides for research, demonstration, or instructional purposes.

Pure Food and Drug Laws

The Federal Pure Food Drug and Cosmetic Act (FPFDA) called the "Pure Food Law" was passed in 1938. Basically, it regulated the use of pesticides on agriculture commodities to prevent excessive amounts of harmful chemical residues in foods, drugs, and cosmetics. Provisions for the establishment of tolerance levels were included.

In June 1955, the Miller Amendment to the Pure Food Law became effective. It is known as the Miller Bill, Public Law 518, Section 408 of the Food, Drug and Cosmetics Act. One of the Miller Bill's main provisions was the determination of the latest safe date to use each pesticide (days to harvest) on a food crop prior to harvest and a tolerance rating for the pesticide on that crop. Three tolerance rating classes were established:

No Tolerance (Exempt).—A material so safe that there is no need to establish a tolerance. Examples:

Allethrin	Sulfur
Pyrethrum	Lime-sulfur
Rotenone	Thuricide
Petroleum Oils	(Bacillus thuringensis)
Bordeaux Mixture	Ryania

Zero Tolerance.—A material so toxic that no residue (0 ppm) is permitted on food crops at market time. Examples:

Dinitros	Hydrocyanic acid
HETP	Mercury compounds
TEPP	Selenates
Heptachlor	Endrin
Aramite	

Specific Tolerance.—A tolerance is given for each pesticide on a specific crop in parts per million (ppm). Examples:

Substance	Ppm	Crop
Terbufos (Counter)	0.05	Corn
Carobofuran (Furadan)	0.1	Corn, grain
Diazinon	0.75	Apples, pears
Carbophenothion (Trition)	0.8	Tree fruits
Methyl parathion	1.0	Small grains
Azinophosmethyl (Guthion)	2.0	Tree fruits
Dimethoate	2.0	Alfalfa

Substance	Ppm	Crop
Chloropropylate (Acaralate)	5.0	Apples, pears
Benomyl (Benlate)	7.0	Apples, pears
Malathion	8.0	Tree fruits, small grains
Carbaryl (Seven)	10.0	Apples, pears
Captan	25.0	Apples, pears
Diazinon	40.0	Alfalfa
Methoxychlor	100.0	Alfalfa

Any amount above tolerance rating on that crop after harvest is unsafe and the crop is condemned unless the excess can be removed before marketing.

The 1958 Amendment to "The Pure Food Law" requires tolerances for pesticide residues in processed foods in addition to the raw agricultural products previously covered and also includes the controversial Delaney Clause which states that any chemical found to be a carcinogen (cancer-causing) when fed to laboratory animals *at any dosage* may not appear in foods consumed by humans. Another amendment in 1964 requires the use of appropriate signal words on the front label of all toxic pesticides, e.g., Danger-Poison, Warning, Caution, and Keep Out of Reach of Children (see Table 12.1)

The Food, Drug and Cosmetic Act of 1938 has undergone amendments and is now administered by the Food and Drug Administration of the Department of Health, Education, and Welfare. As it now stands, the Act:

1. Provides for monitoring of food crops for pesticide residues and enforces tolerances.
2. Establishes food additive (post-harvest application of pesticides) tolerances, monitors them, and prosecutes violators.
3. Provides for working jointly with EPA to register pesticides on animals.
4. Provides for monitoring of pesticide residues in animals by the Meat Inspection Division of the United States Department of Agriculture.

Environmental Protection Agency

The Environmental Protection Agency (EPA) has set up the following regulations with regard to re-entry of fields treated with pesticides:

No unprotected person may be in the field being treated.

No pesticide application is to be permitted that will expose any person to pesticides, either directly or through drift, except those involved in the application.

If labeling for worker re-entry is more restrictive than the general standards specify, the label instructions must be followed instead of the general regulations.

Special re-entry requirements for some pesticides:
Re-entry Waiting Time

48 Hours	24 Hours
Ethyl parathion	Azinophosmethyl (Guthion)
Methyl parathion	Phosalone (Zolone)
Demeton (Systox)	EPN
Monocrotophos (Azodrin)	Ethion
Carbofenothion (Trithion)	
Oxydemetonmethyl (Meta Systox-R)	
Dicrotophos (Bidrin)	
Endrin	

All other pesticide-treated fields may be re-entered without protective clothing after the spray has dried or the dust has settled unless the pesticide is exempted from re-entry requirements.

Appropriate, understandable warnings of pesticide applications are to be given to workers either orally, by posting, or both.

The above regulations are the most important to certified applicators.

Occupational Safety and Health Act (OSHA) of 1970 is administered by the Occupational Safety and Health Administration of the Department of Labor. This law requires any employer with eight or more employees to keep good records of all work-related injuries and illnesses involving medical treatment, loss of consciousness, restriction of work or motion, or transfer to another job, and to make periodic reports. Minor injuries needing only first aid treatment need not be recorded.

It also requires investigation of employee complaints that may be related to pesticide use, re-entry, or accidents.

Regulations governing agricultural aircraft operations are administered by the Federal Aviation Administration of the United States Department of Transportation. It issues commercial and private aircraft operator certificates for such operations under Title 14, Code of Federal Regulations, Part 137.

NEW YORK STATE PESTICIDE REGULATIONS

Most states plan to pass pesticide regulations closely paralleling those of the Federal EPA but some may establish more rigid standards when they think them necessary. Some of the New York State Department of Conservation regulations are offered as an example.

In January 1971, the New York State Department of Environmental Conservation released A, B and C lists of "restricted use pesticides" with instructions and restrictions for each.

A Pesticides may be distributed, sold, purchased, possessed, and used *only* with the issuance of a commercial or purchase "A" permit.

B Pesticides may be distributed, sold, purchased, possessed, and used *only* upon issuance of a "B" commercial permit for very restricted and limited uses.

C Pesticides are all banned for any use in the state of New York.

A Pesticides		B Pesticides	C Pesticides
Actidione	Ethion	Aldrin	Benzene
Aldicarb	Guthion	Arsenicals	hexachloride
Carbofuran	Methylbromide	Chlordane	DDD, TDE
Chloropicrin	Parathion		DDT
Cyanides	Phorate	Dieldrin	Endrin
Dasanit	Sulpho-tepp	Heptachlor	Mercury compounds
Demeton	TEPP	Lindane	Selenates
Dinitros	2,4,5-T	Sodium	Strobane
		fluoroacetate	
Di-syston	Dichlorovos	(1080)	Toxaphene
Endosulfan	Warfarin		Thallium sulfate

When the Federal EPA comes out with its restricted pesticide list, the state will have to combine the A and B lists into one restricted list. The state-restricted list will have to include all pesticides on the Federal list plus any other it deems necessary.

For exact restrictions, see the Section 15 Part 155.2 of the New York State Environmental Conservation Law since there are many special qualifications for use: e.g., above a certain strength, etc.

These lists will be reviewed from time to time and amended to comply with federal regulations or when conditions warrant changes.

Unrestricted Pesticides (Partial List, 1975)

Insecticides		Fungicides	
Malathion	Thuricide	Sulfur	Karathane
Rotenone	Carbaryl (Sevin)	Ferbam	Dodine (Cyprex)
Pyrethrum	Methoxychlor	Zineb	Benomyl (Benlate)
Dursban	Diazinon	Maneb	Coppers
Aspon	Miscible oil	Dyrene	Bordeaux mixture
Dimethoate	Metasystox R		
(Cygon)			
Kelthane	Tedion		
Dipel	(tetradifon)		

Applicators and Applications

1. Pesticides must be applied according to their labeling.
2. All pesticide application equipment that fills from state surface waters must have a check valve or other anti-siphon device to prevent backflow.
3. Presently all commercial applicators must be registered with the Department of Environmental Conservation.
4. In the future all commercial applicators in New York state must be registered whether they use restricted-use or general-use pesticides.
5. When the regulations become fully effective those who are certified will no longer need "A" permits.
6. Commercial applicators must keep records showing pesticide used, dosage, method of application, time, and place for three years. They must be available for inspection.
7. Applicators must apply pesticides carefully so that they do not contaminate crops, pastures, lands, and waters.
8. Applications must not be made under wind conditions causing drift and contamination.
9. Surplus pesticides and containers must be disposed of according to rules and regulations of the State of New York.
10. Only approved refuse disposal sites, sanitary land fills, or incinerators may be used.
11. Small quantities of burnable containers may be burned daily if local public health laws permit it.
12. Properly cleaned, reusable containers can be reused for purposes approved by the Commissioner of the Department of Environmental Conservation. They cannot be reused for food or water containers or for storage of cooking utensils, dishes, or clothing.
13. Surplus pesticides may be buried under at least 18 inches of dirt in such a way that ground or surface water is not contaminated.
14. Permits are required for all uses of pesticides on all surface waters that have an outlet.

LABEL INFORMATION

Abbreviations.—On pesticide labels or used in recommendations:

        WP—wettable powder
        F—flowable
        G—granules or granular
        D—dust
        SP—soluble powder
        EC—emulsifiable concentrate
        SC—spray concentrate

The three most important pieces of advice concerning the labels on pesticide packages are to read them, understand them, and follow them. The information is put there to protect the applicator, the consumer and the environment as well as the crop. The standardized formats of container labels for general use and restricted use pesticides as proposed by the EPA Registration Divison will require the following items of information.

1. *Product name.* The trade or proprietory name to distinguish it from similar products with the same active ingredients marketed by other companies. The common name, if any, may also be given.
2. *Company name and address.*
3. *Net contents.* The weight in English and (or) metric system.
4. *EPA pesticide registration number.*
5. *EPA formulator or manufacturer establishment number.*
6. *Ingredients statement.* The active ingredients (chemical name) in per cent—the actual toxic ingredients that do the killing; and the inert ingredients in per cent—carriers or additives without toxic properties.
7. *Pounds or kilograms of actual toxic ingredient per gallon or liter of liquid concentrate.* This is important in figuring the correct dosage per acre or hectare.
8. *Precautionary statements* (on front panel). Several signal words or symbols based on the toxicity of the product are required by law (see Table 12.1 Categories of Toxicity). If pesticide is in Category I, the signal words are *Danger—Poison* (in red) with a skull and crossbones; in Category II, the signal word is *Warning;* in Category III, the signal word is *Caution;* and in Category IV, the signal word "caution" is not required, but precautionary statements may be necessary. All pesticide labels must bear the statement, "Keep out of reach of children". These statements and others required on the back panel are very important to the safety and health of the applicators.
9. *Statement of practical treatment for accidental exposure or poisoning.*
10. Other precautionary statements (on back panel). Covering such topics as hazards to humans, domestic animals and the environment (bees, birds, fish, etc.); and physical or chemical hazards such as flammability, tendency to explode, skin or eye irritation and protective clothing and equipment.
11. *Re-entry statement* (if applicable). Waiting period after application for safe re-entry into treated fields by workers.
12. *Restricted-use pesticide block.* On front panel if pesticide is re-

stricted in use. Indicates sale and use limited to certified applicators or persons under their direct supervision.

13. *Misuse statement.* Indicates a federal law violation if product is used in a manner inconsistent with its labeling.

14. *Category of applicator.* Indicates in which of the ten categories an applicator must be certified in order to legally use the restricted pesticide. (See FIFRA Act–Certification).

15. *Storage and disposal information.* The rules and regulations of the state must be followed.

16. *Directions for use.* Pesticides are registered for specific crops. It is a violation subject to fine or imprisonment for a pesticide to be used on any other crops not designated on the label. Crop directions provide information on method of application, dosage, timing and days to harvest (the least time interval before harvest that a pesticide may be applied to a crop without the danger of exceeding the residue tolerance).

17. *Warranty statement.* Manufacturer's statement declaring the limit of product guarantee or warranty and the responsibilities of users. It does not mean that the company does not have confidence in its product and refuses to stand back of it. Products put on the market have been thoroughly tested and should perform as stated when directions are followed.

SAFETY PRECAUTIONS FOR APPLICATORS

It is important for applicators to keep in mind that pesticides may enter the body by three ways: (1) oral exposure, eating or drinking the material; (2) dermal exposure, absorption through the skin; and (3) inhalation or respiratory exposure, absorption by breathing. This requires keeping them out of your mouth by avoiding licking your lips, smoking, and eating while making applications; preventing them from contacting your skin by wearing proper protective clothing, gloves, and shoes; and preventing from inhaling them by wearing the proper respiratory devices. Furthermore, applicators must be aware that daily or frequent symptomless exposures to certain pesticides, by any one or more of the above ways, may result in accumulation in the body until symptoms of poisoning appear (e.g., organochlorines, cyclodienes) or may result in cumulative effects until severe symptoms appear (e.g., organophosphates).

All pesticides should be considered poisonous, some more so than others. The following are some of the most important safety precautions when applying them.

1. Avoid the use of highly hazardous pesticides (eg., those in toxicity I and II) when safer ones will do the job.

2. Read the entire label carefully before opening the container, *noting all warnings and precautions.*
3. Apply the pesticides only for the crops and target pests designated on the label. Follow the directions! Use of any pesticide inconsistent with the label is illegal.
4. Avoid absorption of pesticides through the mouth, skin or lungs by wearing rubberized gloves, boots, protective clothing and proper respirators when handling, mixing, pouring and applying those in Categories I and II.
5. Keep out of the way of dust or spray drift. Never smoke, eat, chew or lick your lips when making applications.
6. Shower thoroughly and wash protective clothing daily. Wear clean clothes each day. Change clothes immediately if they become contaminated. Use safety precautions when washing contaminated clothing.
7. Destroy empty containers by breaking or crushing them. Dispose of them in a manner approved by the state. Always store pesticides under lock and key in original containers with the label intact. Keep them out of reach of children and animals.
8. Dispose of left-over spray, if necessary, where there is no danger of run-off into streams or bodies of water and where it will quickly drain into the soil. Never leave puddles!
9. Never use containers or spray equipment that has previously been used for 2,4-D type herbicides. Even traces of weedkiller will severely damage sensitive crops.
10. Avoid spraying insecticides toxic to bees during the pollinating period of crops to protect bees and other pollinators. Notify neighboring beekeepers 24 hr in advance to protect their bees if a toxic spray must be applied.
11. Make sure all employees, especially applicators, are fully informed of all the hazards involved with the pesticides used and the precautions to be taken. Post telephone numbers of the nearest doctor with available prearranged medical services and the nearest Poison Control Center (usually located at a hospital).
12. Make sure the applicators and assisting personnel are familiar with the earliest symptoms of pesticide poisoning, e.g., dizziness, nausea, stomach pain, blurred vision, headache, coughing, faintness, or abnormal reactions. When a symptom or symptoms appear, stop working immediately and seek help from a doctor.

Emergency Pesticide Accident Information

The most important source of information in case of any serious pesticide accident is *CHEMTREC* (Chemical Transportation Emergency

Center) sponsored by the Manufacturing Chemists Association in cooperation with the Pesticide Safety Team Network. Emergency information on all pesticide accidents, poisoning, spills, and clean-ups can be obtained 24 hr a day by calling the toll-free number, *800-424-9300.*

Specific pesticide poisoning information can be obtained by writing or by telephone from:

> National Clearing House for Poison Control Centers
> HEW, Food and Drug Administration
> Bureau of Drugs
> 5401 Westbard Avenue
> Bethesda, Maryland 20016

Protective Clothing and Equipment

The type and amount of protective clothing and equipment required for the operator in a pesticide application should be determined by the following factors:

1. The toxicity, mode of action, formulation and concentration of the pesticide.
2. The degree of exposure.
3. The length of exposure.
4. Extent of skin absorption.

Label instructions under *Precautions* for pesticides in Categories I and II should indicate the kind of protection required by the applicator using the more hazardous pesticides.

Protective Clothing.—*Coveralls* or *Waterproof Rain Coats.*—Coveralls may be worn if they will not be wet through by the mist or spray; otherwise wear a waterproof rain suit.

Head and Neck Coverings.—Waterproof or water repellent hoods are good for protecting the head and neck area. For the head alone waterproof rain hats, safety hard hats or caps are sufficient, but do not use old felt hats or absorbent head gear—they will absorb pesticides in the sweatband area and produce skin contact.

Gloves.—Use natural rubber latex or polyethylene gloves when handling or applying pesticides absorbed through the skin, e.,g., organophosphates, carbamates. Check them frequently for holes. Do not use leather or cotton gloves and do not expose cuts and abrasions in the skin to pesticides.

Boots.—Wear only rubberized boots. Do not wear leather or canvas shoes. Keep boots clean inside and outside.

Goggles.—Protective shields or goggles are necessary when mixing highly poisonous concentrates and when the applicator is subjected to spray or dust drift. Keep the goggles clean (Fig. 14.1D).

Respiratory Devices.—The Occupational Safety and Health Adminis-

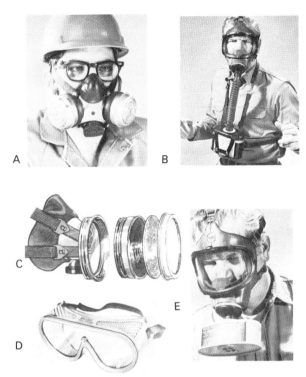

(A, B, D, E, Courtesy Mine Safety Appliances Co;
C, Courtesy Pulmosan Safety Equipment Corp.)

FIG. 14.1. SAFETY EQUIPMENT

A, Dust respirator; B, Industrial fumigant mask; C, Combined dust and fumigant mask;
D, Protective goggles; E, Chin style vapor mask.

tration (OSHA) requires the use of appropriate respirators when hazardous pesticides are being used and also requires that employees be properly instructed in their use, care, and limitations. The chief requirements are:

1. Written standard operating procedures governing the selection and use of respirators shall be established.
2. Respirators shall be selected on the basis of hazards to which the applicator is exposed.
3. They should be cleaned, inspected and disinfected after each day's use and stored in a clean and sanitary location.
4. Only persons physically able to perform the work and use the equipment should be assigned tasks requiring respirators.
5. Only approved or accepted respirators adequate for the particular

hazard for which they are designed should be used, e.g., approved by the U.S. Department of Interior, Bureau of Mines.

There are three types of respiratory devices available for use by commercial and private pesticide applicators: (1) chemical dust or mist respirators; (2) chemical cartridge respirators: and (3) gas mask respirators.

The chemical dust or mist respirators fit over the nose and mouth and are equipped with air filters only (Fig. 14.1A). They are effective only for filtering out pesticidal dusts, mists, aerosols, smokes, and sprays. They are not effective for vapors or gases or those pesticides that have fumigating properties. Filters under heavy exposure should be changed twice a day.

Chemical cartridge type respirators may provide full face coverage, or cover only the nose and mouth and offer no eye protection (Fig. 14.1C,E). In addition to filters to remove dust or spray particles they are equipped with one or two replaceable cartridges containing activated charcoal or other absorbent material. They are generally adequate for outdoor applications of hazardous pesticides. Cartridges should be replaced after each day's use or when odors are detected.

Gas mask-type respirators cover the entire face including the eyes (Fig. 14.1B). They are equipped with the canister directly attached to the mask or attached by a flexible rubber air hose. Canisters contain absorbent material and long-life filters. They are recommended for use in greenhouses and buildings when fumigants or other highly hazardous pesticides are being applied. A safer, supplied air-type respirator with a special compressed air tank attached by an air hose to the hood is recommended when fumigating or applying highly toxic pesticides in confined spaces.

The use of respirators or protective devices does not eliminate the need for other precautions in handling pesticides.

QUESTIONS FOR DISCUSSION

1. What would happen if you attempted to bring in some plants with you on your return from a foreign country?
2. Why are quarantine regulations necessary to control the movement of plant materials into the United States? From state to state?
3. Why are many of our serious insect pests and plant diseases imported ones?
4. Who is responsible for enforcing plant quarantine regulations in the United States? How are they enforced?
5. Why is the Delaney Clause in the 1958 Amendment to "The Pure Food Law" controversial?

6. Why is it important to have specific residue tolerances established for specific crops?
7. Do you feel the requirements for users of "restricted-use" pesticides are justified?
8. What is meant by "days to harvest"? Whom does this provision of the Miller Bill protect?
9. What is the relationship of "days to harvest" to "tolerance" ratings for a pesticide on a specific crop?
10. What determines the "days to harvest" designation for a pesticide? The tolerance rating?
11. What type of pesticides require field re-entry regulations?
12. Why are detailed records of all work-related injuries required of employers?
13. Can a state establish standards for the use and handling of pesticides more rigid than those of the federal government? Less rigid?
14. Discuss the importance of following pesticide label information conscientiously.
15. What pesticide name is apt to mean more to a grower, the chemical name, the common name, or the trade name?
16. Why is it important to act quickly when pesticide poisoning is suspected?
17. What types of pesticides require protective clothing and respirators?
18. Why are liquid absorbent items of clothing such as gloves, hats or shoes dangerous for the operator to wear when applying organophosphates?
19. What type of respirator should one use in indoor fumigation with a highly toxic fumigant? In outdoor spraying with a highly toxic organophosphate?
20. What principle or principles of pest control do each of the pest control techniques discussed in the previous chapters illustrate?

Chapter 15

Application Equipment I

No matter how skillful and experienced an operator is he cannot do a better job of applying pest-control chemicals than the efficiency of the application equipment will let him. Conversely, a poor operator may get discouraging results even with the best equipment. The ideal requirement for equipment for chemical applications is to make possible an even, thorough coverage of the plants or crops at the right time with the minimum of time, labor, and materials.

The application of chemicals for pest control is one of the most disagreeable chores of the average grower. It has undergone great changes from the days of the crude broom or brush for slopping on the materials. In fact, the changes that have taken place have made the standard spray gun obsolete for large-scale operations, since these changes tend to eliminate labor and increase the speed of application without sacrificing evenness and thoroughness of coverage. Depending on their nature, pest-control chemicals may be applied with sprayers, dusters, mist blowers, wet dusters, air sprayers, or aerosol or fog applicators. The equipment may be carried, wheel-borne, tractor-borne, truck-borne, or air-borne.

SPRAYING EQUIPMENT

The purpose of spraying equipment is to eject chemical solutions, suspensions, or emulsions in the form of finely divided droplets so that they may be deposited evenly on the plants or pests. A sprayer consists of definite parts, each with a definite function. Since most sprayers are built on similar principles, an understanding of how one operates will make it easy to understand how all function.

The Parts of a Sprayer and Their Function

The parts common to most large sprayers are the tank to hold the spray, the pump to take up and discharge the liquid under pressure, a relief valve and compression chamber to provide an even pressure, the spray gun or nozzle to break up the liquid into the desired spray, and the engine to operate the mechanism (Fig. 15.1).

The Tank.—This may be made of wood, polyethylene, steel, stainless steel or fiberglass. Tanks vary in capacity from a few to 500 gal., depending on the pump capacity and the purpose for which the sprayer is to be used. Steel tanks are generally protected on the inside with a corrosion-resistant epoxy coating. The tank should contain a good *agitator* to keep materials in suspension and to prevent uneven spray strengths. The tank should also be equipped with a *strainer* through which the spray materials are strained; and a second strainer in the intake pipe leading to the pump, because hard particles are an important source of spray-pump trouble and nozzle clogging.

The High Pressure Spray Pump or Reciprocating Pump.—(Figs. 15.1, 15.2A). This consists of one to four cylinders in which the spray pressure is developed, and corresponding *plungers* or *pistons* that create the spray pressure by drawing the spray into the cylinders through intake valves and forcing it out under pressure through outlet valves. The plungers are made to fit tightly within the cylinders, usually by means of wear resistant cups that can be replaced as the need arises. The number of cylinders and their diameter have a direct bearing on the capacity of the pump. Engine-driven pumps have capacities from 1–2 gal. per min to more than 60 gal. per min, depending on the pressure and on the number of cylinders and their cubic displacement. The pump capacity in turn determines how many nozzles of a certain size may be used and still maintain the desired pressure. Relief and unloader valves are required on high pressure pumps.

The Valves.—These are usually of the ball-and-seat or disc type, are made of bronze or stainless steel, and work on the principle of allowing flow of liquids in one direction only (Fig. 15.2 B, C). Thus each pump cylinder must have two valves—one to allow the spray to come in and the other to let the spray go out. The valves are one of the chief sources of trouble in spray machines and should be easy to reach for cleaning. The following valve troubles may cause loss of pressure or loss of spray volume:

1. Spray or dirt particles may wedge on the seat and prevent the ball or disc from fitting snugly on it.
2. The balls may grow sticky from the spray materials and become stuck on the seats if the machine is put away without a thorough cleaning.
3. The seat or ball may become scratched or dented by hard particles and may allow a backward flow.
4. The ball or seat may become corroded or worn through long use and thus not function properly.

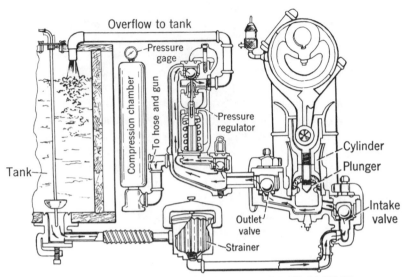

FIG. 15.1. PARTS OF A TYPICAL SPRAYER INDICATING COURSES OF SPRAY

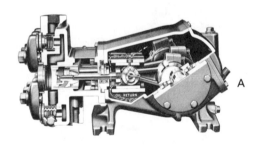

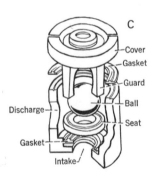

FLAT VALVES — Stainless steel valves, valve springs and valve seats — Positive — Fast acting — Efficient — Less slippage — Longer life — Seat tapered into cylinder casting for easy removal.

(Courtesy F. E. Meyers and Bros. Co.)

FIG. 15.2.

A, High pressure spray pump; B, Stainless steel flat valve; C, Ball and seat valve.

Compression Parts.—From the cylinders the spray material passes to a T-joint, one end of which leads to a hydraulic damper, *air* or *compression chamber.* Where pressure is built up by air compression to produce a cushioning effect and to prevent sudden drops in pressure or a pulsating spray. In one-cylinder pumps, it makes possible an even, continuous flow of spray even between strokes of the piston. If an air chamber is missing, as it may be on some small, hand-operated sprayers, a spray is produced only at each downward stroke of the piston.

The other end of the T-joint leads to a *pressure regulator, relief-valve,* or *unloader valve* (Figs. 15.1, 15.3). The pressure regulator is used for adjusting the pressure to the desired height and in addition functions as a safety valve, preventing excess pressure from being built up, especially when the guns are shut off. When the liquid pressure exceeds that for which the regulator spring was set, the spring is forced to contract. This allows the excess pressure in the form of spray to pass the relief valve and flow back into the tank. Even when the spray guns are opened fully there should be some spray flowing back into the tank, since the pump output should always be greater than the spray gun or nozzle capacity. The unloader valve is a relief valve (60–600 psi range) which contains a separate mechanism which gives the relief valve free by-pass when the pressure build up is greater than the pressure setting. It is normally used with piston pumps, so that when the boom or handgun is not in operation, the pump may run without load on the power source. When the boom is turned on the system pressure automatically returns to the predetermined operating pressure setting. For most spray work, pressures from 300 to 600 lb per sq in. (psi) are common. Small, garden-type engine sprayers are usually run at 150–200 lb pressure and weed sprayers at 30–60 lb pressure.

A *pressure gage* is generally placed somewhere in the line, beyond the spray pump. It may be on the air chamber or near the point of attachment of the spray hose, so that the operator can see that the desired pressure is being maintained.

The hose on sprayers is built to withstand high pressures and varies in length according to the needs of the operator or the spraying conditions. The standard length on orchard sprayers is 50 ft. Each hose on a sprayer leads to a spray gun or a spray boom, depending on the type of crop for which the sprayer is adapted.

Centrifugal Pumps

In a centrifugal pump, liquid enters through the center of the impeller (A). By centrifugal force and the cam action of the vanes (B), the liquid is thrown outward into a spiral passage (C) and then to the outlet port (D).

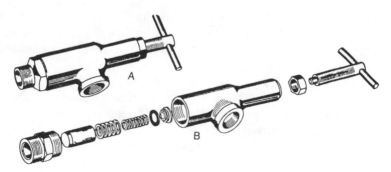

(Courtesy Hanson Equipment Co.)

FIG. 15.3.

A, Pressure relief valve; B, Exploded view.

All energy is imparted to the liquid during its travel from the impeller center to the tip of the vanes. Consequently, pressure remains constant in its travel through the spiral passage to the outlet port. Therefore, to increase discharge pressure and flow rate, the rotational speed must be increased. This is accomplished by a step-up drive from the pump to the tractor PTO. It is not uncommon to increase rotational speeds as high as four times that of a 1000 rpm PTO. Obviously, the higher the speed, the greater the seal wear.

These are high volume, low pressure pumps that depend on rotor speeds up to 4200 revolutions per minute (rpm) to produce up to 120 gal. per minute (gpm) at pressures from 30–200 lb per sq in. (psi) depending on the model (Fig. 15.4A). The dual stage centrifugal pump model will produce pressures up to 200 psi with little drop off at higher spray volumes; whereas pressure drops off rapidly in most other centrifugals at the higher gallonages. The pumps are normally used where abrasives may be present and volume for agitation is required with the full dispensing system in operation, e.g., in most applications of pesticides where the pressures developed are adequate. The pump does not require a relief valve since it will bypass within its rotor body, but it requires a flow valve in the delivery line to regulate the pressure by restriction of the dispensing unit. The pump may or may not be self-priming, should not be run dry and should always be mounted below or near the level of the base of the spray tank if not self priming. It may be run from a tractor power take-off (PTO) or by a gasoline engine.

Roller Pumps

These pumps are semi-positive (intake greater than output) and depend on rotor speeds to produce from 3–20 gpm at 25–200 lb pressure.

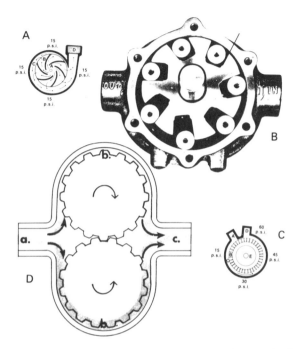

(A, C, Courtesy Delavan Mfg. Co.;
B, Hypro Div. Lear Siegler, Inc.; D, Jabsco Pump Co.)

FIG. 15.4.

A, Centrifugal pump; B, Roller pump; C, Turbine pump; D, Gear
pump.

The pumps will self-prime (draw from the tank a short distance above
the spray source) and require a relief valve ahead of the dispensing
shut-off or control valve (Fig. 15.5). They are not designed for abrasives
and must not be run dry. They may be powered from the power take-off
(PTO) of a tractor or by a gas engine. Nylon or Nylatron rollers are
recommended for pesticides, especially for wettable powders; rubber
rollers are for pumping water or wettable powders only. The pumps
operate normally at 500–1200 revolutions per minute (rpm).

The pumps are designed with a case that is eccentric to the shaft and
rotor. The rotor has 6–8 equally-spaced slots, depending on the model,
containing precision rollers which act as vanes to force liquid through
the annular gap between the rotor and eccentric housing. Since hydraulic and centrifugal forces keep the rollers in contact with the hous-

ing, displacement is semi-positive and capacity is roughly proportional to pump speed, particularly at lower pressures.

Turbine Pumps

In conventional turbine pumps (Fig. 15.4C) liquid enters the suction port (A), is caught by the vanes (B) on the outer end of the impeller and thrown outward into the channel (C). The configuration of the channel forces the liquid back to the vane. This circulation of the liquid from vane to channel to vane repeats many times during the passage of liquid from the inlet port to the outlet port (D). Each cycle of liquid imparts additional energy. Pressure is increased as the liquid passes through the pump. In theory, a turbine pump would be ideal for agricultural spraying applications where the pump can be mounted directly to a 1000 rpm PTO. However, when pumping wettable powders or liquid fertilizers, the suspended materials will accumulate around the shaft and between the impeller and pump housing (E). This will cause the pump to wear excessively or malfunction. The problem is corrected in the much more efficient and durable Delavan Turbo 90 model. The Turbo 90 is a high volume, multi-purpose turbine pump. It will pump abrasive wettable powders and liquid fertilizers, and is perfect for high volume, low pressure spray rig pumping applications. A by-pass line or relief valve is normally required.

Epoxy Plastic Gear Pumps

Gear pumps are self-priming and nearly positive in action when unworn (See Fig. 15.4D). As the gear teeth separate on the inlet side of the pump, the spray from inlet (a) fills the spaces between them. The liquid is then carried around the periphery of the bore (b) between the gear teeth. The spray is squeezed out of the discharge (c) as the teeth mesh. This gives a smooth positive flow rate that is directly proportional to the rpm of the pump. The pumps will produce up to 12 gpm at pressures up to 100 psi at 1750 rpm. They are powered from a tractor power take-off (PTO) or a gasoline engine. The gear pump requires a relief valve on the line.

Spray Guns

A spray gun consists of a long or short barrel made of light metal, one end of which is attached to the hose and has a shutoff valve, the other end of which has a nozzle for producing the type of spray desired. The shutoff may be of the trigger type, or it may be of the twist type (Fig. 15.6).

The small-volume spray guns generally used with hand-operated equipment commonly have an adjustable nozzle and a trigger shutoff. All-purpose spray guns produce a short, misty spray to a long, coarse spray, depending on how far the shutoff valve is opened. This type of spray gun is useful where trees and shrubs of various heights have to be sprayed at the same time. The orchard or broom type of spray gun (Fig. 15.6) may have anywhere from two to eight nozzles arranged on a crosspiece so as to throw a long, fairly flat spray. This type of gun allows for speed along with thorough coverage as long as the trees are not too high. The shade-tree gun (Fig. 15.6) is usually provided with a long barrel and a rifle-like nozzle that throws a coarse spray to great heights, breaking up as it ascends. This type of gun is used in combination with high-pressure, high-output pumps to spray tall shade trees.

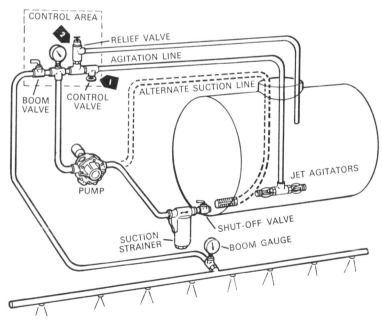

(Courtesy Hypro Div. Lear Siegler, Inc.)

FIG. 15.5. ROLLER PUMP SPRAYER INSTALLATION

Spray Booms

Spray booms are used for row crops and are of various designs, determined by the crops to be treated (Fig. 15.7). They may be made to take care of as many as sixteen or more rows at a time. For insect and disease control their nozzles should be so arranged as to spray the plants from all sides—usually three nozzles per row are used. Wide spray booms up to 84 ft are made with hinged wings so that they can be folded out of the way of fences, gates and other obstacles. Some are hydraulically operated.

Spray Nozzle Parts and Their Function

The parts of a typical spray nozzle are shown in Fig. 15.8. They are the base, the cap, the disc, the washer, the vortex plate, and the strainer. In large spray guns the strainer may be left out of the nozzle and the vortex plate may have more openings.

Clogging is the most frequent cause of trouble with nozzles, and

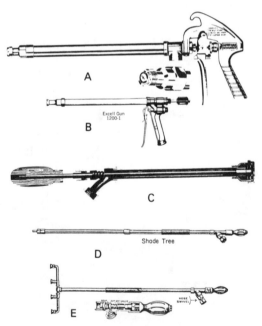

(A, B, Courtesy Spraying Systems Co.;
C, Courtesy Friend Mfg. Corp; D, E, F. E. Meyers and Bros. Co.)

FIG. 15.6. SPRAY GUNS

A, B, C, All-purpose spray guns; D, Shade tree gun; E, Orchard or broom gun.

(*Courtesy of F. E. Meyers and Bros. Co.*)

FIG. 15.7. AN 8-ROW SPRAY BOOM IN ACTION ON A POTATO FIELD

anyone who uses them should be able to take them apart and put them together again correctly. The base of the nozzle and the cap hold all the parts in place. The disc is equipped with one hole in the middle and is usually numbered from 2–10. The number denotes the diameter of the orifice in $1/64$ in. The size of the orifice regulates the volume of spray and in hand-operated equipment indirectly affects the pressure that can be built up. Disc holes may enlarge or may become lopsided after much use, and the discs may need to be replaced. In nozzles producing a flat spray pattern the orifice is in the center of a long, longitudinal groove in the disc.

The washer under the disc prevents leakage and produces an eddy chamber, the depth of which in turn determines the angle of the spray cone in nozzles producing hollow-cone sprays. The vortex plate has near its outer edge two or more slots or holes that slant through at an angle and throw the spray into a spin in the eddy chamber. This causes the spray to break up into the form of a hollow cone when it leaves the disc. The vortex plate is absent in nozzles producing flat spray patterns. The strainer helps prevent nozzle clogging.

Hand-Operated Sprayers

Atomizer Type.—These sprayers are made to hold from a pint to a gallon of spray and are most useful in and around the home where limited areas must occasionally be treated. This type of sprayer produces a fine mist and may be used with either solutions or suspensions. With suspensions, frequent shaking of the contents is necessary to prevent settling (Fig. 15.9).

Advantages.—Low priced and handy with small quantities of spray on limited areas.

Disadvantages.—Limited to small jobs. Not suitable for good coverage of low-growing crops. Short carrying distance of spray.

Compressed-Air Type.—These sprayers (Fig. 15.9) are made to hold from 1–4 gal. of spray and operate on air compressed in the tank by means of an air pump. These sprayers should not be filled more than three-quarters full of spray, since an air space is necessary to build up air compression. The sprayers are carried slung over the shoulder by means of a strap. They are usually equipped with a trigger shutoff type of spray gun operated by one hand. They are useful in small gardens and where a few trees and shrubs of limited height are to be sprayed.

Advantages.—Low priced, and simple to operate.

Disadvantages.—

1. Uncomfortable to carry and operate for any length of time.
2. Must be frequently shaken when suspensions are used.
3. Varying pressure—tank must be pumped up several times before it is emptied.
4. Limited carrying distance of spray.

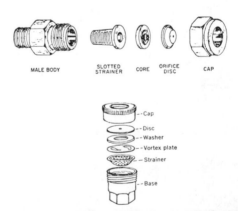

FIG. 15.8. THE PARTS OF TWO STANDARD SPRAY NOZZLES

5. Parts wear out quickly and corrode easily unless made of stainless steel.

Knapsack Sprayers.—These hold 2–5 gal. of spray and in general have the same uses as compressed-air sprayers. They are carried on the back like a knapsack, and the right hand operates a pump equipped with an air chamber to produce a continuous pressure. The left hand operates the spray gun. (Fig. 15.9).

Advantages.—Uniform pressure may be maintained. Equipped with an agitator.

Disadvantages.—Tiring and uncomfortable to operate for any length of time; heavy load; pumping action produces bruises. More expensive than compressed-air sprayers.

Bucket Pumps.—These are small pumps (Fig. 15.9) that may come equipped with or without a bucket that may vary in capacity from 2–4 gal. They are generally equipped with an air chamber so as to produce a continuous spray pressure. Agitation is usually produced by a brass plate connected to the pump handle by a rod, or by a spray jet from the bottom of the pump. They are used for many of the same purposes as the other two types of sprayers. Two people are generally required to do a good job with this type of outfit: one to carry the bucket and pump, the other to do the spraying. At least 10 ft of hose are needed for increased maneuverability.

Advantages.—Inexpensive; easy to operate and to maintain even, high pressure; and greater carrying distance of spray.

Disadvantages.—Pump and bucket with spray must be carried from place to place, and equipment must be set down to pump. Constant pumping required to maintain pressure. Frequent refillings required for any amount of spraying.

Wheelbarrow and Barrel Sprayers.—These types of sprayers (Fig. 15.9) are equipped with 10–50 gal. tanks in which are placed pumps similar to bucket pumps but larger and more powerful, so that greater spray volume and pressures may be used. The whole outfit is usually mounted on a wheelbarrow frame or on a 2–4 wheeled frame for easy portability. A compression chamber is essential with these pumps to supply a continuous spray at an even pressure. These sprayers are made in a number of styles and designs and are adaptable for shrubs, small orchards, and fairly large vegetable gardens. Those for greenhouses are generally built quite narrow so as to fit easily in the walks between the benches.

Advantages.—All the advantages of the bucket pump plus portability. Higher pressures and spray volumes obtainable than with bucket

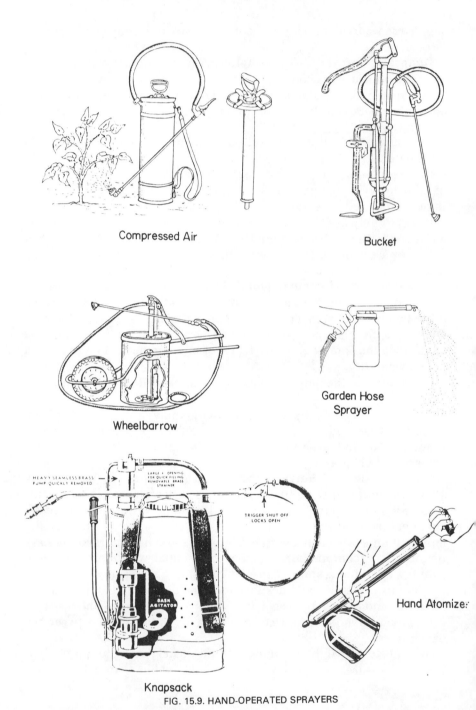

Compressed Air

Bucket

Wheelbarrow

Garden Hose
Sprayer

Knapsack

Hand Atomizer

FIG. 15.9. HAND-OPERATED SPRAYERS

pumps. Greater spray-carrying capacity, and can take care of greater areas than any of the previously mentioned types.

Disadvantages.—Two men required for effective operation, and pushing or pulling full tank is hard work, especially on soft ground. Limited to fairly level areas; ineffective use of manpower for the amount of work done.

Power Sprayers

Engine-operated sprayers vary from small, handy, wheelbarrow or two-wheel mounted sprayers with a 10–15-gal. tank capacity to huge 500-gal. sprayers mounted on four-wheel or truck chassis.

Small Power Sprayers.—These may be of the wheelbarrow type, barrel type, or estate type (Fig. 15.10). They vary greatly in construction and design, but essentially they all function in a similar manner. Those suitable for greenhouse spraying are generally equipped with electric motors. The chief defect of the one-wheeled wheelbarrow types is that they are easily upset and thus subject to damage. The two-wheeled types are more stable, but those with tank capacities of 25–50 gal. are difficult to push or pull around when they are full of spray. Some of these may have a tractor hitch and obviate the necessity of manpower for locomotion. These sprayers are usually equipped with 1–2 cylinder pumps with capacities of 1–6 gal. per min and with maximum pressure limited to 200–300 lb. A good agitator is essential to keep materials in suspension. They are usually equipped with a hose and spray gun, but some may be adapted for row-crop spraying by attaching a spray boom. These

FIG. 15.10. SMALL POWER SPRAYERS

sprayers are suitable for flower and vegetable gardens, greenhouses, shrubbery, and small orchards.

Advantages.—Labor of pumping eliminated. Can be operated effectively by one man. Maintains high, even pressure. Pumps have greater capacity than hand-operated ones, and perform the job more quickly and efficiently.

Disadvantage.—Much more expensive than hand-operated equipment.

Spraying Attachments on Tractors.—Many tractors may be equipped with detachable spraying equipment. These spraying attachments usually consist of a small pump, a spray tank, and a spray boom or hose and spray gun. The power takeoff for the pump generally consists of a belt and pulley arrangement with the tractor's engine. In most cases low-pressure, low-volume pumps are used with the power takeoff (Fig. 15.5).

Spraying attachments on garden tractors generally perform as well as similar, unattached spray units and have the advantage of mechanical mobility. They have the same uses as separate engine-powered sprayers of similar pump and tank capacities.

Advantages.—Engine can be used to operate attachments other than sprayers. Cost of spraying attachments may be less than cost of similar spray units. Motive power all mechanical. Better adapted for row crop or ground spraying than equipment pulled or pushed by man.

Disadvantages.—Mounting and taking off spray equipment takes time and labor—equipment often must be taken off when tractor is used for purposes other than spraying. Good spray agitation may be lacking.

Power Sprayers.—Large power sprayers may be of several types. Their tank capacities range from 100–500 gal. They may be mounted on a 4–6 wheel chassis and have their own engines, or they may be of a trailer type mounted on two wheels with their own engines or with a tractor power take-off to run the pump. The 4–6 wheel cut-under models can be turned in their own length. The pump capacities vary from 12 to more than 60 gal. per minute, with engines of suitable horsepower to run them. Maximum spray pressures run from 400 to more than 800 psi. They may be equipped with one or more 50 ft lengths of hose, with spray guns, with a spray boom for row-crop spraying, with spray masts, or with turrets of nozzles for rapid coverage of fruit trees. The tractor takeoff models eliminate the cost of an extra engine, but they require tractors of a certain design and horsepower to work effectively (Fig. 15.11).

Advantages.—Give good service over a period of years. Cover large areas of row crops in good time (up to 35 acres per day). Spray can be

(*Courtesy of John Bean Mfg. Co.*)

FIG. 15.11. A TRACTOR-TRAILER SPRAYER IN OPERATION

directed with accuracy. Adaptable for many types of crops and for all insecticides and fungicides.

Disadvantages.—Require large amounts of water—must have good source of water. Equipment is heavy, especially when tank is filled, and may bog down in wet weather. Require heavy tractor equipment for motive power. Spray gun applications slow, hard, distasteful work as compared to use of spray masts or new speed application equipment. Expensive in labor, material, equipment, and time when compared to latest equipment.

Concentrate Sprayers.—Concentrate sprayers may be variously known as air sprayers, speed sprayers or mist blowers. They all make use of an air current to carry concentrated pesticidal solutions, emulsions or suspensions into the foliage of tree- or row-crops and to apply these evenly at the proper dosage level. Some use a high-velocity, low volume air current; others use a large volume, low velocity air current as the pesticide carrier.

A typical concentrate sprayer (Fig. 15.12) consists of a small hydraulic high pressure or centrifugal pump to move the concentrate mixture from

(Courtesy of F. E. Meyers and Bros. Co.)

FIG. 15.12. A CONCENTRATE AIR SPRAYER IN ACTION IN AN ORCHARD

the tank to the discharge opening of a large blower. There, as the spray leaves the nozzles, it is atomized by the shearing action of the air from the blower and carried as a fine spray through the trees or crops. The pump and the blower are usually powered by one large engine. The most efficient concentrate sprayers require a large volume of air at velocities up to 90 mph as a carrier for the spray for adequate coverage. They may use high pressure, centrifugal or roller pumps depending on the needs of the grower.

The discharge head of the blower, inverted teardrop or circular in shape, is designed to discharge from only one side or both sides of the sprayer as it moves between the rows of trees or crops. Some models, e.g., those for shade tree applications, have an adjustable discharge head mounted on a turn-table platform. The inverted teardrop discharge head permits a greater and faster moving volume of air from its upper area for a long reaching swath and a less forceful discharge from its lower area for safe coverage of close-in rows.

Field crop sprayers may have one-way discharge heads that will produce spray swaths up to 60 ft wide or two-way outlets (one from each

(Courtesy of F. E. Meyers and Bros. Co.)

FIG. 15.13. A CONCENTRATE AIR SPRAYER IN ACTION IN A POTATO FIELD

side of the machine) that will produce combined swaths up to 120 ft wide in winds up to 10 mph (Fig. 15.13). One-sided tree sprayers with air volume to 95,000 cfm can cover tree heights up to 70 ft. Two-sided orchard sprayers with the same air volume will generally cover tree heights up to 30 ft.

Advantages.—Reduction in labor costs of as much as 60–70%. Saves time, water and water haulage. Run-off eliminated—saving in spray materials. Less wheel damage to row crops—drive rows further apart. Faster application rate makes it possible to complete applications during optimum time for spraying (mornings or evenings). Operator less exposed to the pesticides. Gives a more thorough, even coverage. Blower may be used for other purposes, e.g., frost protection, crop drying, nut harvesting, cotton defoliation.

Disadvantages.—Pesticides must be applied with great accuracy and under favorable weather conditions. Operator must be well trained and conscientious. Rate of travel determines coverage. Greater danger of spray injury to fruit or foliage because concentrates are used. Spray pattern at higher concentrations tends to lose visibility—more difficult to judge proper application. Concentrate spray unit more costly than high pressure hydraulic unit.

QUESTIONS FOR DISCUSSION

1. Why can application equipment be no more efficient than the man operating it?
2. What are the different ways in which chemicals may be applied for pest control?
3. What are the comparative merits of wood, steel, stainless steel, polyethylene and fiberglass spray tanks?
4. Why is a good agitator so important in a sprayer?
5. What are the effects of a clogged strainer in a sprayer?
6. Explain how a high pressure spray pump works.
7. What is the importance of a relief valve in a sprayer?
8. What happens when the nozzle output exceeds the pump capacity? Vice versa?
9. What are the possible causes of a drop in pressure or of insufficient pressure?
10. What may go wrong with ball or disc valves in a spray pump?
11. Of what importance is a compression chamber?
12. Where are all the possible trouble points on a sprayer located?
13. Explain why an orchard type of spray gun is most effective for trees that are not too high.
14. What are the features of a desirable type of spray boom for row crops?
15. What are the advantages of concentrate spraying over normal dilute spraying? Disadvantages?
16. What part of an apple tree normally has the most wormy apples or diseased foliage? Why?
17. What is the function of each of the parts of a spray nozzle?
18. What are the advantages and disadvantages of the different spray pumps?
19. What are important considerations in purchasing a sprayer?
20. Why do sprayers able to apply up to 30X concentrates offer little practical advantage over those applying up to 10X concentrates?
21. What type of sprayer would be most suitable for a small home garden? A small fruit planting? A commercial fruit orchard? A commercial row crop? A huge alfalfa field?
22. Under what conditions are "spray rings" practical?
23. Which would be more advisable to buy: sprayer attachments for a small garden tractor that one already owns or a separate, engine-driven, small spray unit?
24. If you were a big fruit or field crop grower would you switch to concentrate spraying equipment? Why?

Chapter 16

Application Equipment II

DUSTERS

Dusting equipment makes use of an air stream to carry chemicals in finely divided and dry form onto the plants.

The Parts of a Duster and Their Function

Dusters are much simpler in construction than sprayers and have relatively few working parts (Fig. 16.1). The typical engine-driven duster is composed of a dust hopper holding from 25–200 lb of dust, with an agitator to keep the dust from caking and also to keep it flowing evenly through an adjustable opening in the bottom of the hopper that

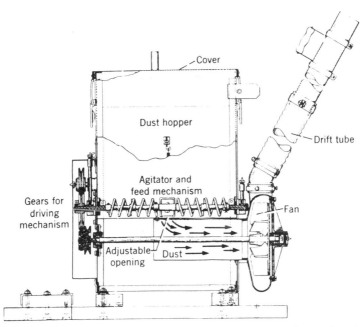

(Courtesy of Niagara Chemical Division, Food Machinery Corp.)

FIG. 16.1. CROSS SECTION OF A DUSTER SHOWING WORKING MECHANISM

305

connects with the blower unit. The fan-type blower unit creates a powerful air current that picks up the dust as it drops, either in the air chamber or directly against the blower fan blades, and carries it out through the distributor head and delivery tube or tubes. The distributor head may be equipped with baffles to distribute the dust-laden air evenly through all the delivery tubes. For orchard or tree dusting, a single, large flexible drift tube or an adjustable fish-tail type of delivery tube is generally used. For low growing trees, multiple, fixed-type drift tubes may be employed. For row-crop dusting, multiple delivery tubes—2–18 or more on a dust boom—are used. From 1–3 are needed per row, depending on the type of coverage desired (Fig. 16.2).

The nozzles at the end of the delivery tubes are generally of a fish-tail design so as to spread the dust evenly. They may be open type or closed type nozzles. Delivery tubes and nozzles on row-crop dusters are adjustable to fit rows and plants of various widths and heights. A hood-like enclosure over the nozzles will increase dust deposit, reduce drift, and allow effective dusting, even with some breeze blowing.

Small Hand Dusters.—These may be of the telescope type and consist essentially of one cardboard cylinder within another; the bellows type; or the plunger type (Fig. 16.3A). In all these types an air stream is produced that picks up the dust and drives it out onto the plants. The dust containers in these dusters may hold from 1 pt to 2 qt of dust, but they should never be filled more than two-thirds full, because the dust may pour out or come out in too heavy a stream. These dusters are valuable in very small gardens where their prolonged use is not necessary.

Advantages.—Cheap, and simple to operate. Speedier, lighter and easier to use than sprayer.

Disadvantages.—Continuous pumping tiresome. Good only for small areas, and dust supply uneven.

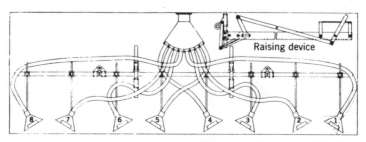

(Courtesy of John Bean Mfg. Co.)

FIG. 16.2. A DUST BOOM SHOWING NOZZLES ARRANGED FOR THE PROPER COVERAGE OF FOUR ROWS

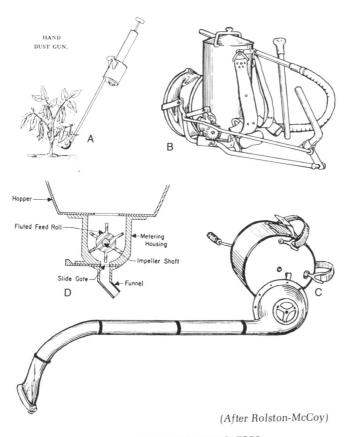

(After Rolston-McCoy)

FIG. 16.3. DRY PESTICIDE APPLICATORS

A, Hand duster; B, Knapsack bellows duster; C, Rotary crank
duster; D, Metering assembly of a granular distributor.

Knapsack Dusters.—These are of two types: the bellows type and the
rotary fan type. The bellows type (Fig. 16.3B) is excellent for spot
dusting of individual plants or hills. It produces a cloud of dust at each
constriction of the bellows. The long discharge tube with fish-tail nozzle
is useful in dusting low growing plants as well as small shrubs and trees.
These dusters hold from 5–10 lb of dust and are made of light material so
their total weight is not more than about 25 lb when filled.

The rotary-fan duster operated by means of a crank is preferable for
row-crop dusting (Fig. 16.3C). It produces a steady, even cloud of dust
when the crank is turned at an even speed while the operator is walking

along the row. The delivery tube and nozzle are adjustable to different heights and directions. Double nozzles may be used to cover two rows at a time. Much greater areas may be covered with this type of duster, but it may not give as good coverage as the bellows type unless care is taken to regulate the height and direction of the nozzle so as to blow the dust into the plants from the side instead of from above. Both the above dusters may be quite tiring for several hours of continuous use. The rotary-fan duster is also designed for dusting from mule or horse-back.

Advantages.—Even flow of dust; they have a positive dust-feeding device and an adjustable opening in dust hopper. They can carry enough dust to cover large area, and are simple to use. Rapid application.

Disadvantages.—Much more costly than hand-carried equipment. Rather tiring for continuous operation.

Granule Distributors.—These are similar to fertilizer distributors which either broadcast the granular formulation or feed it in a band over the seed furrow during or after planting. The equipment consists of a hopper, a metering device with a rotating agitator for metering the granules through an adjustable orifice, and a spreading device to produce a uniform band (Fig. 16.3D).

Advantages.—Equipment can be used in combination with the seed planter. Dust hazard eliminated, and dosages can be accurately controlled. Equipment relatively inexpensive.

Disadvantages.—Use limited to granular formulations to be used as systemics or to soil pests. Granular formulations may not be available. Material must be applied with precision; nature and size of granules may vary the application rate. Calibration of the equipment necessary for each formulation applied. Proper speed for the calibration must be maintained.

Power Dusters.—These range from small, knapsack types with engine motive power to powerful row-crop or tree dusters (Fig. 16.4) pulled by tractor. Some dusters are of the trailer type and operate by power takeoff from the tractor that pulls them. The large dusters generate an even, high-velocity dust cloud and are suitable for protecting large acreages.

Advantages.—Suitable for large acreages. No water required, so no time wasted in refilling. Speedy—saves labor and time; less trouble—not as many things to go wrong as in a sprayer. Equipment light and heavy tractor equipment not necessary. Cheaper to purchase than sprayers that can do the same job.

Disadvantages.—Use limited to suitable weather conditions, especially for tree dusting. May not give as good control of some insects or plant diseases because of rapid weathering of chemicals. Dusts may carry to other crops and injure them or leave dangerous residues.

(*Courtesy of John Bean Div., FMC*)

FIG. 16.4. A TREE DUSTER IN ACTION IN AN APPLE ORCHARD

AIRPLANE AND HELICOPTER APPLICATION OF CHEMICALS

Custom pest control by means of airplanes or helicopters has gained in popularity (Figs. 16.5 and 16.6). It is especially suitable for large acreages and inaccessible areas, since it offers an extremely rapid method of application, with coverage of hundreds of acres of crops or forested areas possible in a single day. Dusts, concentrated sprays, or fogs may be applied from the air. For crops, airplane or helicopter dusting has been found more effective than spraying because of better and more even coverage. The down-draft from the wings of the plane or rotors of the helicopter tends to drive the dust down in a swirling cloud so that some coverage on the undersurface of foliage is also obtained. This is not true of sprays, which cover only the top of the foliage. Fog or concentrated mist application of the new organics from planes or helicopters is very effective for space control of flying pests of man and animals and of foliage-chewing insects. It is not suitable for applications of fungicides, because the residual effect is slight and thoroughness of coverage is poor. As little as a pint of material per acre may be applied from the air as a fog or mist.

Airplane application is faster than helicopter application, but the

(Courtesy of Bell Aircraft Corp.)

FIG. 16.5. A HELICOPTER DUSTING A FORAGE CROP

FIG. 16.6. HELICOPTER SPRAYER

helicopter can go into fields and wooded areas too dangerous for planes to tackle. The helicopter also has the advantage of being able to land for reloading in any small clearing close to the scene of operations, whereas the airplane must have a regular landing field, often miles away from the areas to be treated.

Advantages.—Large areas may be treated in a comparatively short time (up to 300 acres in an hour). Grower eliminates cost and upkeep of equipment, and use of labor. No crop damage from equipment—the wheels of ground equipment often cause considerable damage to crops. Areas inaccessible to ground equipment or where use of ground equipment is impracticable may be treated.

Disadvantages.—Applications dependent on weather conditions— planes or helicopters may be grounded by weather conditions, or wind may be too strong for effective application. Grower must depend on contracting party to make applications when needed—this is not always possible. Control usually not as good as with ground equipment, and pesticide drift may present hazards to the environment.

AEROSOL OR FOG APPLICATION EQUIPMENT

Liquified gas aerosol generators consist of special metal or.steel containers ("bombs") fitted with a valve and a nozzle or with high-pressure hose and special spray gun and nozzle. There are no working parts: one has merely to open the valve to release the contents of the container in the form of an aerosol or a fog. The carrier is a liquified gas such as "Freon" or methyl chloride in which the active ingredients are either dissolved or emulsified. This liquified gas produces the pressure and atomizing effect on release into the air through a suitable nozzle. Some "bombs" should be carried inverted when discharging the aerosol.

Other aerosol generators (Figs. 16.7 and 16.8) may produce thermal aerosol smokes, thermal aerosol fogs, steam aerosol fogs or cold aerosols. (See Pesticide Formulations—Aerosols, for methods of their production.) Under outdoor conditions the generators depend mainly on natural air currents to drift the aerosol over the crops or through the trees to be protected. Some mist sprayers that use an air blast may very closely approach aerosol atomization.

Aerosols are very useful for insect and red spider control in enclosed places. Under outdoor conditions they have been used for insect control on canning crops, field crops, vegetables, and in forested areas. They have also been valuable for space control of flies, mosquitoes, and gnats. Aerosol application has been successful on row-crop insects when the aerosol is confined temporarily over the plants by a canopy or a light tarpaulin.

(Courtesy of Bes-kill Corp.)

FIG. 16.7. A FOG GENERATOR IN USE FOR MOSQUITO CONTROL

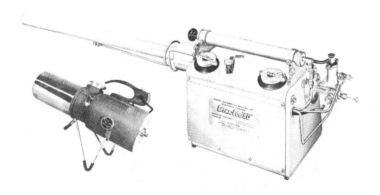

(Courtesy Curtis Automotive Devices, Inc.)

FIG. 16.8. HAND CARRIED FOG GENERATORS

Aerosols that have a fairly rapid settling rate are best for outdoor use on plants, since they are not easily carried away by air currents and leave some residual effect on the foliage. Their use in orchards is limited, because air movements interfere with complete coverage, and very little residual deposit is obtained. No effective fungicidal aerosols have been produced, mainly because complete, even coverage and a residual deposit is necessary for disease control.

Advantages.—Very effective for pest control in enclosed places— disperses into inaccessible areas. Liquified gas aerosol application requires no expensive equipment. Fog generators relatively inexpensive. Rapid, easy treatment of enclosed spaces—takes 5 min to treat a greenhouse of 100,000 cu ft capacity. Very small quantities of toxic ingredients can be dispersed over large areas. Creates no residue problem.

Disadvantages.—Residual effect slight. Containers must be returned for refills when used up. More expensive in materials cost than other application methods. Not effective for disease control. Outdoor applications limited by weather conditions. Difficult to confine to a definite area outdoors. Use limited to special situations—not practical for field crops.

SOIL FUMIGATION EQUIPMENT

The application of soil fumigants over large acreages has been greatly speeded up and improved in recent years. The hand-carried and operated injectors are suitable for small areas such as greenhouses, cold frames, seedbeds, and small gardens, but are impracticable for larger areas. The new applicators are similar in design to fertilizer applicators

(Courtesy of Mack's, Caldwell, Idaho)

FIG. 16.9. A TRAILER TYPE OF FUMIGATOR SUITABLE FOR LARGE AREAS

or seed planters: the smaller ones may be hand-pushed, whereas the larger ones that treat a swath several feet wide are tractor-borne and are capable of treating many acres a day (Fig. 16.9). The fumigant is piped into the bottom of furrows made by sharp, cultivator-like tines, and sometimes a roller is attached to press the soil down. The fumigant in the form of a liquid or granular material is fed into the furrow at a regulated rate. The rate of flow is adjusted to the speed of the equipment to give the required dosage of the fumigant per acre.

QUESTIONS FOR DISCUSSION

1. What are the advantages of dusting over spraying? Vice versa?
2. What are the limitations of dusting for crop pest control?
3. Do you consider the present, small, hand-operated sprayers and dusters on the market models of perfection?
4. Which is to be preferred for a small home vegetable garden: a bellows knapsack duster or a rotary fan duster? For a large vegetable garden?
5. What were the chief faults of traction powered sprayers and dusters?
6. Which will give the better coverage of heavy-foliaged plants: spraying or dusting?
7. What advantages has aerial application of chemicals over ground application? Disadvantages?
8. Explain how a liquid aerosol generator works? A thermal fog generator? A steam aerosol fog generator?
9. Why do chemicals in aerosol form tend to be more effective for certain pests than the same ones in spray or dust form?
10. Why has aerosol application been adopted so readily by greenhouse operators? By pest control operators?
11. Why don't aerosols disperse like fumigants?
12. Why is soil fumigation becoming more and more practical?
13. How useful is a fumigation chamber to a greenhouse operator? To a nursery stock grower?
14. If your garden soil was heavily infested with nematodes, would you prefer to apply an effective soil fumigant or an equally effective systemic nematocide? Why?

Chapter 17

Selection, Care, and Manipulation of Application Equipment

SELECTION OF APPLICATION EQUIPMENT

In purchasing application equipment the following factors should be considered:

1. The comparative advantages and disadvantages of the different forms in which chemicals may be applied to plants. Chemicals for pest control may be applied to plants as sprays, dusts, granulars, air-driven mists, spray-dust combinations, and aerosols. These are discussed in the previous chapters in connection with application equipment. The form best suited to the crop and its protection should be selected.

2. The acreage to be covered or the time allowable for one complete application. The equipment should allow for one complete application in the maximum of 5 days but preferably in 1–2 days. This will allow for better control through proper timing or through more frequent applications when necessary.

3. The type of crop to be treated. The machine and its discharge apparatus should be adapted for the most efficient and thorough coverage of the crop.

4. Its efficiency in crop protection. This can be determined only by actual experimental trials. The state experiment stations usually carry on such trials and are in a position to advise on relative efficiency. The experience and advice of growers that have used similar equipment in previous seasons is also of great value.

5. The initial cost of the equipment in relation to the returns from the crop to be protected. Under ordinary conditions, new application equipment should pay for itself in the maximum of 5 years, from the standpoint of the added efficiency and crop protection it gives. Where fields are small, individual growers may not be

315

justified in purchasing expensive equipment; it may be more economical for a group to form a "spray ring" and buy a sprayer or hire a commercial applicator, but this should not be at the expense of improper timing.

6. The work and skill involved in operating the equipment. The number of men required to operate the equipment and the skill required of the operators to do a good job should be considered.

7. Cost of operating the equipment. This involves the cost of the operators, the fuel, and the replacement of parts that tend to wear out.

8. Availability of important parts for inspection, removal, and replacement, This is of primary importance, for when a breakdown occurs much time may be lost or the equipment may be laid up at a critical time for preventing or controlling an outbreak of insects or disease.

9. Pump capacity. The spray pump should be able to supply sufficient volume (gpm) and maintain pressure when all nozzles are open. High pressure pumps with the minimum of 16 gpm capacity will generally handle effectively two spray guns with ordinary nozzles. Spray pumps for row crops should be able to supply, from several nozzles at high pressures, 100–125 gal. of spray per acre at 4 mph, whereas those used in weed control should be able to supply as little as 10 gal. per acre or less at low pressures. Dusters should generally be able to apply from 30–50 lb of dust per acre at 4 mph.

10. Operating pressure. For most efficient coverage and penetration, high pressure sprayers should be able to operate at pressures from 300–600 psi. Some may operate up to 800 or more psi, but in most cases such high pressures are unnecessary and may put a severe strain on the engine.

11. Tank capacity or dust hopper capacity. More time is lost in refilling sprayers than in any other sprayer operation. The tank or hopper capacity should be maximum from the standpoint of saving time in refilling, but at the same time it should be adaptable to the field conditions and the motive power available. Too heavy a spray load that causes bogging down in wet fields is just as inefficient as too light a spray load necessitating frequent refills.

12. Durability of the equipment. Experience of other growers with the equipment should be investigated, as well as the length of time the company has been in business and its reputation. Application equipment should give good service for at least 5 years without any major repairs or replacements.

CARE OF EQUIPMENT

The proper care of application equipment will add many years to its useful life and will help to avoid serious breakdowns that may interfere with the proper timing of applications. The following practices will help keep equipment in good running order:

1. Lubricate thoroughly all working parts before each day's work. Check especially the lubrication of the cylinder and piston assemblies, since these parts usually get the hardest wear.
2. Always add materials through the strainer in the spray tank; this will help prevent valve trouble and nozzle clogging.
3. At the end of each day's use wash the tank carefully and run clear water through the pump, hose, and spray nozzles for a minute or two. This is done to prevent corrosion of parts and valve sticking when the sprayer is used again.
4. Clean outside parts of sprayer of spray residue that may produce corrosion.
5. At the beginning of the season check piston packing, gaskets, and valves. Replace worn out parts if necessary.
6. Drag the hose as little as possible. If hose is being dragged, keep alert to the possibility of snagging it on an obstruction and snapping it off. See that hose does not rub against the wheels when it is wound up. Many a good hose has been ruined by being ground against a wheel.
7. At the end of the season clean sprayer thoroughly: drain all water from the engine, pump, and tank; run a 50% solution of Prestone through the complete sprayer; lubricate thoroughly all external working parts, spray gun and nozzle parts; drain oil from engine and pump and replace with new oil; and store sprayer without removing the Prestone.
8. Dusting equipment, when being used or being prepared for storage, in general, should receive about the same treatment for lubrication of working parts and protection from corrosion as spraying equipment.
9. Use a separate sprayer for 2,4-D and similar weed-killers to avoid crop damage. Even traces can be injurious.

SPRAYING AND DUSTING TECHNIQUES

The effectiveness of application equipment is based on the skill of its operators. The same piece of equipment may give good, fair, or poor control of pests as determined by its operator's ability to do a thorough job at the right time. There is no better way to learn the correct techniques of application than through watching, and practicing continuously under the guidance of an experienced operator. Only after considerable practice can one be relied on to know what good coverage is and to be able to attain it.

Spray Patterns and Their Adjustment

The nozzle has an important bearing on the effectiveness of the whole spraying operation. Almost any type of desired spray may be obtained by using the right nozzle or by adjusting the nozzle. The four types of spray patterns that may be produced by spray nozzles are: hollow-cone spray, solid-cone spray, flat spray, and solid-jet spray.

The hollow-cone type of spray (Fig. 17.1A) is the one most often used in close-up spraying. The angle of the cone may be made either narrower or wider by varying the depth of the eddy chamber. A thicker washer will increase the depth of the eddy chamber and in turn give a narrower spray cone. Too wide a spray-cone angle reduces carrying distance and spraying efficiency. The solid-cone spray (Fig. 17.1B) is produced by boring a hole somewhat smaller than the disc hole in the center of the vortex plate. This type of spray tends to increase carrying distance and spray coverage and is useful in applying herbicides to the bark and cut surfaces of woody perennials.

The flat or fan-shaped spray most useful in weed-control work is produced by a nozzle without a vortex plate and with a grooved disc. The solid-jet spray is produced by a nozzle without a vortex plate. It is useful for long-distance spray coverage such as reaching the tops of tall trees. All-purpose types of spray guns have a twist type of control valve that increases the depth of the eddy chamber as it is opened. This makes possible all types of sprays from misty, hollow cone with a narrowing angle to solid jet.

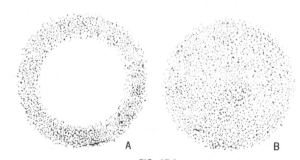

FIG. 17.1.
A, HOLLOW-CONE SPRAY PATTERN; B, SOLID-CONE SPRAY PATTERN

Spray Adjustments

The type of spray produced, its carrying power, its volume, and its fineness may be adjusted to some extent by various manipulations of the spray pump and the spray nozzles.

Spray volume is increased by increasing the pressure, the size of the disc orifice, and/or the number of nozzles.

Spray fineness is increased by increasing the pressure. In hand-operated equipment, this is done by pumping faster or by using a disc with a smaller orifice. In power equipment, this is done by increasing the tension on the spring in the pressure regulator.

Carrying distance of spray is increased by:

1. Increasing spray pressure. In the case of hollow- or solid-cone sprays, pressures beyond 800 psi actually tend to decrease carrying power, because the spray is broken up into so fine a mist that it does not have the body to carry as far as a coarser spray under lower pressures.
2. Reducing the angle of the spray cone by increasing depth of the eddy chamber.
3. Removing the vortex plate. This produces a solid-stream spray. Solid-stream sprays as produced by shade-tree guns continue to give increased carrying distance beyond 800 pounds pressure.
4. Using a larger disc orifice, but at the same time maintaining or increasing the previous pump pressure.

Techniques in Spraying Plants

Amount to Apply.—Determine the best type of spray boom, spray mast, or spray gun and nozzles to use for the job. Keep in mind that an even, thorough coverage is required with the minimum of waste. A light dripping immediately after an application or light wetting in concentrate spraying, indicates good coverage without excessive use of spray, and heavy dripping means too heavy an application and waste. The speed of movement of the spray rig or that of the spray gun governs the amount of spray applied.

Tree-Spraying Techniques

1. In spraying a tree from the ground with a spray gun, spray the tree first from underneath, if necessary; then start on the outside at the point opposite the sprayer and work around the tree towards the sprayer so as not to end up with the hose wrapped around the trunk. When the tree is completely sprayed, the operator should be able to move directly to the starting point on the next tree (Fig. 17.2).

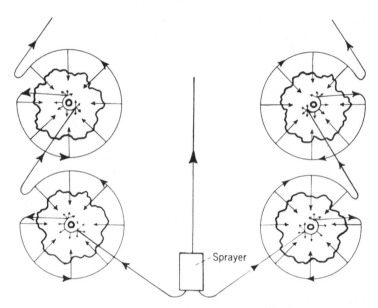

FIG. 17.2. PROCEDURE IN SPRAYING TREES FROM THE GROUND BOTH FROM UNDERNEATH AND FROM THE OUTSIDE

Arrows extending from lines indicate direction of spray.

2. When spraying the outside of a tree, treat it as a solid object and the spray gun like a paint brush. Move the spray gun at a steady speed in vertical up and down sweeps, overlapping slightly each time to obtain complete coverage (Fig. 17.3).

3. Hold the spray gun a moment at the top end of each sweep and flick it slightly if necessary to get good top coverage.

4. As the spray gun is brought down, back away a step or two, if necessary, to avoid sticking the end of the spray gun into the tips of the lower branches and thus miss covering the tips and possibly causing injury to foliage or fruit.

5. The speed with which the spray gun is moved will determine the coverage. Gauge the speed so that the foliage is wetted and a light dripping results.

6. It is not wise to attempt to spray when there is a strong breeze (as indicated by a flag standing straight out); but if an application is essential, a spray may be put on from the windward side only. Never spray directly into the wind; most of the material will be wasted, and the operator will be covered with it.

7. Take note of the wind direction. When the point where the spray gun begins to face into the wind is reached, cover as much of the

tree as possible from that angle, shut off the spray gun, shift to the same position on the opposite side of the tree, face the last sprayed portion, and continue application around the tree (Fig. 17.4).

Spraying Trees While Riding.—Two methods may be used in spraying trees with hand guns with the operators mounted on the equipment. In the first method one man stands on top of the tank on a special platform and covers the top part of each tree on one side as the spray rig moves by. The other man stands on a low platform mounted behind the spray tank and covers the lower part of each tree. In this operation the spray rig moves steadily from tree to tree with an increase in operation speed, but with some loss in thoroughness of coverage (Fig. 17.5).

In the second method, applicable only to trees that are fairly close together, both men stand on the platform on top of the tank. Each one covers half of each tree on his side of the spray rig as it moves by. This method still further increases the speed of application but usually with additional loss of thoroughness.

Concentrate Spraying

Concentrate spraying involves the distribution of a small quantity of active pesticidal ingredient accurately and evenly over a large area. Normally a pesticide is diluted with water as the carrier at the usual recommended dosage per acre or per tree and applied with a high pressure sprayer to the point of run-off. This is designated as a 1X application. Concentrate spraying is the application of a pesticide on

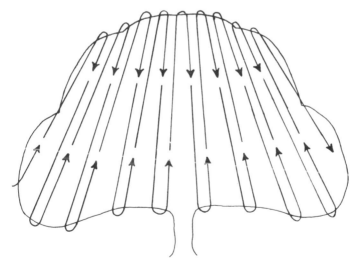

FIG. 17.3. DIRECTION OF SPRAY-GUN MOVEMENTS IN SPRAYING A TREE
FROM THE OUTSIDE

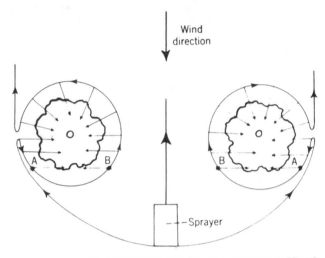

FIG. 17.4. PROCEDURE IN SPRAYING IN ORDER TO COVER THE TREE AND
NOT OPERATOR IN BREEZY WEATHER

A, Shut gun off and walk to point B; B, Turn gun on, point gun in
direction indicated by arrows.

(Courtesy of A. B. Farquhar Co.)

FIG. 17.5. A METHOD OF TREE SPRAYING THAT INCREASES THE SPEED OF
COVERAGE WITH SOME LOSS IN THOROUGHNESS

foliage only to the point of wetting, also at the usual recommended dosage per acre or per tree but substituting air as the carrier for much of the water. In this way, the same coverage can be obtained with concentrates from 2–10X the normal dosage with ½–¹/₁₀ as many gallons of spray. For example, if a tree requires 20 gal. of dilute (1X) spray, it will require 4 gal. (2 gal. on each side) of a 5X concentrate and only 2 gal. of a 10X concentrate; or if a crop requires 200 gal. per acre of a dilute (1X) spray, it will require only 50 gal. per acre of a 4X concentrate. For most concentrate spraying, concentrations of 2–10X with high volume air at velocities up to 90 mph are employed. Machines with concentration capacities up to 30X with low volume, high velocity air streams are available, but offer no significant advantages. Lower spray concentrations are more tolerant of human, mechanical, and natural errors with less chance of fruit and foliage damage. Adequate coverage with increased spray concentration requires increased carrying air volume, not increased air speed. Air outlet velocities higher than 80 mph may cause crop blow down as well as foliage and fruit damage. Also as spray concentration is increased and the volume applied reduced, smaller droplet size is necessary for adequate distribution, and coverage of tree tops or swath ends becomes more difficult because very fine droplets do not impinge readily on surfaces. To get greater spray atomization so that higher concentrations can be applied evenly, pump pressure must be increased. Dormant oil sprays no higher than 3–4X concentration at ¹/₅ normal dilute gallonage should be used to avoid tree injury.

With an adequate sprayer for the tree size or swath width, a driving speed of 2 mph with a 4–5X spray concentrate is generally satisfactory. Where foliage is exceedingly dense, as in citrus trees, a rate as low as 0.8–1 mph is common practice in order to fill the trees with spray laden air. To get heavier coverage of tree tops or foliage the pace must be slowed. For row crops the spacing of the drive rows should allow for a little overlapping of the swath ends (Figs. 15.12, 15.13).

Pointers on Dusting

Dusting is most successful when the foliage is wet and the air is calm. Such conditions are usually most prevalent in the evening or early morning. In tree dusting, the drift tube is so manipulated that the dust cloud completely envelopes the tree while the duster moves along at a steady pace. The tree dusters that have single or double fish-tail discharge heads that are adjustable while dusting are satisfactory mainly for small-sized trees. The latest models of dusters that use large volumes of air at low velocity and have a precision-type feed into the air stream tend to give better and more uniform distribution of the dust.

Calculating Dosages on a Per Acre Basis

The calculating of dosages on a per acre basis is especially important in row-crop spraying and weed spraying with herbicides. The correct amount of each chemical should be applied to the plants on a per acre basis and not at the recommended dilution of the chemicals on a per 100-gallon basis which may be rendered inaccurate by the pressure, speed of travel and size of the disc holes. It is important to be able to check on the gallons per acre that a piece of equipment puts out so that the correct amounts of chemical may be applied to the plants by changing the concentration in the spray tank.

Method for High-Gallonage Applications

The following method is most applicable to high-gallonage row-crop or ground-covering equipment:
1. Start the spray motor, be sure everything is working properly and that nozzles are not clogged.
2. Fill the tank completely with solution of approximately the proper concentration.
3. Open the valve to the spray boom.
4. Drive 40 rods at normal tractor speed (approximately 3–4 mph).
5. Close the valve at the end of 40 rods and measure the number of gallons used by measuring the amount needed to fill the tank completely again.
6. Multiply the number of gallons used by 66.
7. Divide this figure by the width in feet of the spray boom.
8. The answer is the number of gallons of solution that the machine will apply per acre, providing spray pressure and speed of travel are maintained.
9. The right amount of chemical to be applied per acre is then added to each unit number of gallons that is applied per acre. For example: if the sprayer holds 300 gallons, if its output at a certain speed and pressure is 150 gallons per acre, and if 2 lb actual methoxychlor per acre are to be applied, 8 lb 50% methoxychlor (wettable) in a full spray tank should be added.

Method for Low-Gallonage Applications

The following formula is used with low-gallonage, low-pressure equipment:

$$GPA = \frac{GPH \text{ per nozzle}}{0.01 \times mph \times inches\ spacing}$$

where

GPA = gallons per acre;
GPH = gallons per hour per nozzle = cc per
minute × 0.016 or ounces per minute × 0.47;

MPH = miles per hour (3–4 is usual tractor
rate)
Inches spacing = inches between nozzles (measure with ruler).

To test a given piece of low-gallonage equipment to be used with definite-sized nozzle-disc holes at a constant, set pressure:
1. Turn sprayer on.
2. Measure quantity of liquid delivered from 1 nozzle in 1 min either in cc or in tenths of ounces (graduate should not be more than 1 in. diameter for the sake of accuracy and should be made of plastic to avoid breaking).
3. Obtain value for GPH as shown above.
4. Substitute known factors in formula and calculate the GPA.
5. Where any of the three factors is known, the fourth can be calculated from the above formula.

Example: If by the use of the above formula it is determined that the sprayer delivers 10 gal. spray per acre and 1 lb actual pesticide per acre is to be used, 1 lb actual active ingredient should be added for each unit of 10 gal. of water that the spray tank holds.

Method of Diluting Dust or Spray Concentrates to Correct Strength for Application

Almost everyone working with pest-control chemicals has occasion to make definite dilutions of dusts or sprays, starting with the concentrated materials containing a known percentage of the active ingredients. The procedure for determining how much of the concentrated material should be mixed with how much of the diluting ingredient to make a definite amount of insecticide or fungicide of the required strength is as follows:
1. Determine how much of the actual active ingredient is needed in the amount of diluted product desired.
2. Determine how much of the active ingredient is present in 1 unit (lb, gal., cc, or gr) of the concentrated product.
3. Divide the amount present in 1 unit of the concentrated product into the amount needed in the diluted product to determine the number of units of the concentrated product that will have to be mixed with the diluting agent to obtain the desired dilution.

4. The amount of diluting agent required is determined by subtracting the amount of concentrated material needed from the total amount of diluted product desired.

Example.—Make 100 pounds of a 0.75 per cent rotenone dust from a 5 per cent rotenone powder concentrate using pyrophyllite as the diluting agent.

Solution.—100 × 0.0075 or 0.75 lb of actual rotenone will be needed in the 100 lb diluted dust. Each pound of the concentrated powder contains 0.05 × 1 or 0.05 lb of rotenone; therefore, 0.75 ÷ 0.05 or 15 lb of the 5% rotenone powder is needed for mixing with 100 minus 15 or 85 lb of pyrophyllite to produce 100 lb of a 0.75% dust.

The Timing of Fruit Sprays

Since the timing of sprays because of weather and climatic conditions varies greatly in different areas of the country and even in different areas of a state, growers should follow application schedules distributed by their own local agricultural agencies. However, to understand an application schedule, the stages of plant growth that may be referred to therein should be clearly understood by every grower.

Dormant Spray.—In the late winter or early spring before the buds begin to swell.

Green-Tip Spray (Fig. 17.6).—When bud tips begin to show green. Some dormant spray materials may still be applied at this stage.

Delayed Dormant Spray.—When green leaves are from ¼ to ½ inch out. "Superior" oils may still be used; otherwise use summer-spray materials (Fig. 17.6)

Pre-Pink Spray or Cluster-Bud Spray.—When flower buds are beginning to separate from the cluster.

Pink Spray.—When flower buds show pink or other petal color (Fig. 17.7).

Calyx or Petal-Fall Spray.—When about three-fourths of the petals have dropped.

Shuck Spray.—Stone fruits only, when the dried-up blossoms remains (shucks) are splitting and falling off the small fruits (Fig. 17.8).

First Cover Spray.—About 1 week to 10 days after the calyx or shuck spray.

Other Cover Sprays.—As many as five at definite intervals (usually 10-day to 2-week intervals), depending on the insects or diseases present.

Special Considerations in Making Applications.—The application of poisonous chemicals on fruits and seed crops should be avoided during

FIG. 17.6. DELAYED DORMANT STAGE OF APPLE GROWTH

FIG. 17.7. LATE PINK STAGE OF APPLE GROWTH

FIG. 17.8. LATE SHUCK STAGE OF PEACH GROWTH, BEST TIME FOR
SHUCK SPRAY APPLICATION

the blossoming period so as not to harm bees and other insects that are of great value in the cross-pollination of the crops.

In most spray or dust schedules for fruits and vegetables, combinations of insecticides or of insecticides and fungicides save time and effort in the control of the several different types of insects as well as of diseases to which the crops may be subject. From the standpoint of plant injury and effectiveness it is important to know which combinations are compatible and which are not. Table 13.2 lists the compatibilities of some of the more commonly used insecticides and fungicides.

QUESTIONS FOR DISCUSSION

1. In what form are chemicals most effectively applied to plants, taking into consideration coverage and retention?
2. What are the advantages of using equipment that allows complete coverage of an orchard or field in a day or two?
3. What might growers do if individually they cannot afford to purchase an efficient sprayer, or if their fields are too small to warrant the purchase of such equipment?
4. How does one go about determining the efficiency of a machine?
5. Can an unskilled operator do a good job with modern spray equipment?
6. Should a grower be able to inspect, remove, and replace parts in his

application equipment himself, or should he use the services of a qualified mechanic as we do with our automobiles?

7. Of what significance is pump capacity in a sprayer?
8. One gets decreased carrying distance with pressure beyond 800 psi with the standard hollow-cone nozzle. Why?
9. What should determine the tank or hopper capacity desirable in a sprayer or duster?
10. Discuss the usefulness of each of the four types of spray patterns that may be produced by using the proper nozzle.
11. How should one go about narrowing the angle of the spray cone on a standard nozzle?
12. Would using a smaller disc orifice without any increase in pressure produce a finer spray?
13. What type of spray is obtained when a vortex plate is removed from a standard nozzle?
14. Why should a sprayer be cleaned carefully after each day's work?
15. What are the dangers of using 2,4-D in a sprayer also employed for other purposes?
16. What factors should be considered while a tree is being sprayed?
17. Explain the techniques of spraying a tree from the ground with the wind from the south to avoid being covered with spray.
18. Why do very few growers spray trees from the ground?
19. What are the limitations of two-way air sprayers in tree spraying?
20. Why must greater care be used in applying concentrated sprays from mist or air sprayers than in applying diluted sprays from standard sprayers?
21. Under what conditions might dusting be as effective as spraying?
22. In your opinion, what type of equipment can be relied on to give the best coverage of orchard trees? Row crops? Grain or forage crops? Forested areas?
23. What are the limitations of spray schedules?
24. If dripping from the foliage occurs in concentrate spraying, how should that be corrected?

Addendum

330

Pesticides—Trade and Common Names

Insecticides, Acaricides, Nematocides, Mulluscicides

Trade Name	Common Name
Acaraben	chlorobenzilate
Acaralate	chloropropylate
Ambush	permethrin
Animert	tetrasul
Anthio	formothion
Aspon	none
Baygon	propoxur
Baymix	coumaphos
Baytex	fenthion
Bidrin	dicrotophos
Biotrol	*Bacillus thuringiensis*
Biotrol/VHZ	nuclear polyhedrosis viruses
Brofene	bromophos
Bromofume	ethylene dibromide
Brozone	methyl bromide
Carboxide	ethylene oxide
Carzol SP	formetanate
Chlorocide	chlorbenside
Ciodrin	crotoxyphos
CO-RAL	coumaphos
Counter	terbufos
Cyanogas	calcium cyanide
Cygon	dimethoate
Dasanit	fensulfothion
DDD	TDE
DDVP	dichlorvos
De-Fend	dimethoate
Delnav	dioxathion
Dibrom	naled

Di-Captan	dicapthon
Dinitrol	dinitrocresol
Dipel	*Bacillus thuringiensis*
Dipterex	trichlorfon
Di-Syston	disulfoton
DN 289	dinoseb
Doom	*Bacillus popilliae*
Dorlone	D-D mixture
Dowfume MC-2, MC-33	methyl bromide
Dowfume W-85, W-40	ethylene dibromide
Dry-Die, Drione	silica aerogels
Dursban	chlorpyriphos
Du-Ter	fenton hydroxide
Dyfonate	fonofos
Dylox	trichlorfon
Elgetol	dinitrocresol
Elgetol 318	dinoseb
Ethodan	ethion
Famfos	famphur
Fumazone	dibromochloropropane
Furadan	carbofuran
Fundal	chlordimeform
Galecron	chlordimeform
Gardona	tetrachlorvinphos
Guthion	azinphosmethyl
Isotox	lindane
Hexa-nema	dichlorofenthion
Japidemic	*Bacillus popilliae* (milky disease)
Karathane	dinocap
Kelthane	dicofol
Korlan	ronnel
Krenite	dinitrocresol
Kryocide	cryolite
Lannate	methomyl
Larvacide	chloropicrin
Lorsban	chlorpyriphos
Marlate	methoxychlor
Mesurol	methiocarb
Meta-Systox-R	oxydemeton methyl

Mitox	chlorobenside
Mocap	ethoprop
Monitor	methamidophos
Moristan	oxythioquinox
Morocide	binapacryl
Mylone	DMTT
Neguvon	trichlorfon
Nemacur	phenamiphos
Nemafene	D-D mixture
Nemagon	dibromochloropropane
Nemex	D-D mixture
Neopynamin	tetramethrin
Nexion	bromophos
Nudrin	methomyl
Ormite	comite
Orthene	acephate
Otirane	ethylene oxide
Oxirane	ethylene oxide
Penncap-M	methyl parathion
Penphene	dibromochloropropane
Penta (PCP)	pentachlorophenol
Pestmaster	ethylene dibromide
Phosdrin	mevinphos
Phosphine	aluminum phosphide
Phostoxin	aluminum phosphide
Picfume	chlorpicrin
Plictran	tricyclohexytin
Pirimor	pirimicarb
Pounce	permethrin
Profos	ethoprop
Proxol	trichlorfon
Rabon	tetrachlorvinphos
Resistox	coumaphos
Rogor	dimethoate
Rothane	TDE
Ruelene	crufomate
Sevin	carbaryl
Spectracide	diazinon
Sayfos	menazon
Supona	chlorfenvinphos
Systox	demeton
Tedion	tetradifon

Telone	D-D mixture
Temik	aldicarb
Terracur	fensulfothion
Tetron	TEPP
Thimet	phorate
Thiodan	endosulfan
Thuricide	*Bacillus thuringiensis*
Trifenson	fenson
Tri-clor	chloropicrin
Trithion	carbofenothion
Trolene	ronnel
Vapam	SMDC
Vapona	dichlorvos
Vapotone	TEPP
Vendex	fenbutatin oxide
Vidden-D	D-D mixture
Vikane	sulfuryl fluoride
Viron/H	nuclear polyhedrosis virus
Vorlex	MIT
Vydate	oxamyl
Warbex	famphur
Zectran	mexacarbate
Zolone	phosalone

Fungicides–Trade and Common Names

Trade Name	Common Name
Actidione	cycloheximide
Agrimycin	streptomycin
Agri-strep	streptomycin
Antracol	propineb
Arasan	thiram
Basic-cop	low soluble copper
Benlate	benomyl
Bordo	Bordeaux mixture
Botec	dichloran
Botran	dichloran
Bravo	chlorothalonil
Caddy	cadmium chloride
Cadminate	cadmium succinate
Cad-Trete	cadmium chloride

Captan Seed Treater	captan
Carbamate	ferbam
Cyprex	dodine
Daconil	chlorothalonil
Demosan	chloroneb
Dexon	fenaminosulf
Difolitan	captafol
Dikar	mancozeb
Dithane D-14	nabam
Dithane M-22	maneb
Dithane M-45	mancozeb
Dithane Z-78	zineb
Dyrene	anilazine
Exotherm	chlorothalonil
Fore	mancozeb
Formalin	formaldehyde
Formol	formaldehyde
Fermate	ferbam
Fungo	methyl thiophanate
Granox Liquid	carboxin
Kleen	methyl thiophanate
Kocide	low-soluble copper
Kromad	cadmium sebacate
Manzate	maneb
Manzate 200	mancozeb
Mertect	thiabendazole
Microcop	low-soluble copper
Orthocide	captan
Ortho Cop 53	low-soluble copper
Parzate	zineb
Phaltan	folpet
Phytomycin	streptomycin
Polyram	metiram
Scotts No. III	anilazine
Scotts Systemic	methyl thiophanate
Spergon	chloronil
Spot	methyl thiophanate
Termil	chlorothalonil

Terrachlor	pentachloronitrobenzene
Tersan	thiram
Tersan 1990	benomyl
Tersan SP	chloroneb
Thylate	thiram
Tobaz	thiabendazole
Topsin	ethyl thiophanate
Vitavax	carboxin

Glossary

Acaricide. Any chemical or agent used to destroy mites and ticks.

Acervulus. A cushion-like mass of conidiophores and conidia.

Active ingredient. The actual toxic agent present in a formulation.

Adfrontal area. A pair of narrow oblique sclerites bordering the frons on the head of lepidopterous larvae.

Aerosol. A suspension of solid or liquid particles in air.

Alkaloids. Nitrogenous alkaline compounds found in certain plants. Many have toxic properties.

Alkylating agents. Highly active compounds that replace hydrogen atoms with alkyl groups in cells undergoing division.

Alternate host. The other species of plant that is necessary for the completion of the life cycle of some insects and plant disease-producing organisms.

Ametabola. Insects without a metamorphosis.

Amphids. A pair of lateral nerve-connected organs opening on or near the lip region of nematodes.

Anabiosis. The state of suspended animation in some nematodes initiated by desiccation.

Analogues. A series of organic compounds similar in function but not identical in structure to the original compound.

Animal. Any organism that belongs to the animal kingdom.

Antennae (sing. antenna). Paired, sensory, segmented structures found on each side of the heads of insects and of some related forms.

Antimetabolites. Chemicals similar to metabolites but detrimental to their function.

Appendage. Any structure attached to a part of the body of an organism.

Ascospores. Spores produced in an ascus.

Ascus (pl. *asci*). A sac-like container of ascospores, generally eight.

Asexual reproduction. Vegetative reproduction from plant parts other than seeds, or from spores produced by simple budding as in the imperfect stage of a fungus.

Autocide. The use of a pest species for its own destruction.

Autoecious. Completing life cycle on only one host. Term used with rusts.

Bacteria. Microscopic one-celled plants without chlorophyll.

Bactericide. Any substance that destroys bacteria.

Basidiospores. Spores produced on a basidium.

Basidium. A spore-producing, club-shaped hypha, usually giving rise to four basidiospores.

Biomagnification. The progressive build-up of a pesticide residue in the bodies of organisms of a food chain.

Biological control. Control of pests by means of predators, parasites, and disease-producing organisms.

Blight. A disease that causes the rapid discoloration and death of leaves, flowers, twigs, etc.

Brood. Offspring of approximately the same age arising from a single species of animal. Generally used with insects or birds.

Bursa. Lateral cuticular fin-like extensions at or near the posterior end of some male nematodes.

Callus. Tissue overgrowth around a wound or canker.

Canker. A dead, sometimes sunken or cracked area surrounded by living tissues on a stem.

Capitulum. The headlike mouthparts of ticks.

Caterpillar. The "worm-like" stage (larva) of a moth or butterfly.

Cephalothorax. The fused head and thorax found in forms such as spiders and lobsters.

Cerci. Slender, paired and segmented appendages arising from the tenth abdominal segment of some insects.

Chemotherapy. Cure or prevention of disease in plants by internal chemical treatments.

Chemtrec. A toll-free, long distance telephone service that provides 24-hr emergency pesticide information. (800-424-9300)

Chitin. A nitrogenous polysaccharide occurring in the cuticle of arthropods.

Chlamydospore. A thick-walled, asexual resting spore.

Chlorosis. Yellowing of normally green tissues because of the partial failure of chlorophyll to develop.

Cholinesterase. An enzyme in the blood essential for proper nerve function.

Chronic toxicity. The capability of a pesticide to be injurious after repeated exposures. Same as cumulative toxicity.

Circadian rhythm. State of activity of an organism over a 24 hour span.

Cloaca. The terminal chamber of the digestive and reproduction systems in male nematodes before emptying through the anus.

Cocoon. A protective sac, spun by the larvae of many insects, in which they pass the pupal stage.

Compatible. Refers to chemical materials that can be mixed together without changing their effects adversely on pests or plants.

Conidia. Asexual spores produced on conidiophores other than sporangiospores and chlamydospores.

Conidiophore. Simple or branched hypha on which conidia are produced.

Contact poison. A chemical that kills when it contacts some external part of a pest.

Control. Prevention of, or reduction of, losses from pests by any method.

Cornicles. Two tubes on the upper posterior end of the abdomens of aphids, which secrete a waxy liquid.

Corpus. The anterior portion of the esophagus of a nematode usually consisting of a slender anterior portion (procorpus) and a swollen basal bulb (metacorpus or median bulb).

Coxa (*pl. coxae*). The basal segment of an insect's leg.

Crochets. Row of tiny hooks at the end of prolegs on lepidopterous larvae.

Ctenidia. Comblike rows of stout bristles on some fleas.

Cumulative toxicity. Same as chronic toxicity.

Cyst. The dry cuticle of dead adult females of the genus *Heterodera* that protects the eggs within.

Damping-off. Seed decay in the soil or the decay of the bases of seedlings.

Diagnosis. Identification of the nature and cause of a plant trouble.

Diapause. A period of arrested activity in insects in response to seasonal low temperatures or high temperatures and drought.

Didelphic. With two ovaries, in nematodes.

Die-back. Progressive death of twigs or leaves beginning at tips.

Dikaryon. A pair of haploid nuclei that occur in one cell prior to Karyogamy in fungi.

Diluent. Any liquid or powdered material that is used to reduce the concentration of a chemical for spraying or dusting purposes.

Dioecious. Plants that have only staminate or pistillate flowers on one plant.

Disease. Any change in a plant that affects its normal structure, functions, or economic value.

Disinfectant. An agent that kills or inactivates organisms before they attack plants, as in the soil or on seed surfaces.

Dissemination. The transport of inoculum or pest from a diseased to a healthy plant.

Dormancy. A state of arrested activity.

Dorsal. Referring to the back or upper side of an organism.

Ecdysone. Hormone secreted by insects essential to the process of growth.

Ecosystem. The interacting system comprised of all the living organisms in an area and their nonliving environment.

Ectoparasite. A parasite feeding externally on host tissue.

Elytra (sing. *elytron*). The hard, upper wings of beetles.

Emulsifiable liquid. A liquid that will form an emulsion when it is mixed with water.

Emulsion. A dispersion of one liquid in another in the form of tiny droplets.

Endoparasite. A parasite entering and feeding internally on host tissue.

Enphytotic. The sudden and destructive development of a plant disease, usually over large areas (adjective form: *epiphytotic*).

Esophagus. That portion of the digestive tract between the stoma and the digestive system in nematodes.

Estivation. The period of suspended animation in many tropical insects and in some animals in response to hot, dry conditions.

Etiolation. Excessive spindliness in plants, owing to lack of sufficient light or disease.

Etiology. Study of the cause of disease.

Exudate. Liquid discharge from diseased or injured tissues.

Facets. The visual units that compose a compound eye.

Femur (pl. *femora*). The third and usually the heaviest segment of an insect's leg, counting from the body.

Food chain. Populations of organisms in an ecosystem where each population is preyed on by one higher in the developmental scale.

Form. The resting place of a rabbit or hare.

Frass. The wet or dry sawdust-like excrement of borers, usually evident at their exit holes in fruits and other plant parts.

Frons. A triangular plate on the face of insects just above the clypeus.

Fruiting body. A fungous structure that contains or bears spores.

Fumigant. Any material that when in the form of a gas or vapor destroys organisms.

Fungicide. Any agent that kills or inhibits fungi (or, when used in a broad sense, bacteria).

Fungus (pl. *fungi*). A plant without chlorophyll, usually possessing a thread-like body and reproducing by means of sexual or asexual spores.

Furcula. Forked springing apparatus on the ventral side of the abdomen of springtails.

Gall. A more-or-less spherical outgrowth of unorganized plant cells produced by the irritation of bacteria, fungi, mites, or insects.

Gametes. Reproductive cells of fungi with haploid nuclei capable of fusing to form diploid zygotes.

Gametophyte. The asexual or imperfect phase in the life cycle of a fungus.

Generation. Referring to a group of progeny originating from adults of a preceding group.

Germ tube. The hyphal thread produced by a germinating fungous spore.

Gradual metamorphosis. The growth changes in insects from newly hatched nymphs to adults.

Gram-negative. Not capable of being stained by the standard Gram stain.

Gram-positive. Capable of being stained by the standard Gram stain.

Grub. The larva of a beetle.

Halteres. Short-knobbed appendages present in place of the hind wings of true flies.

Haustorium (pl. *haustoria*). A specialized food-absorbing organ that grows into a cell from the hyphae of certain species of fungi.

Hazard. The chance that harm will come to beneficial forms of life from the use of a pesticide.

Hemelytra (sing. *hemelytron*). The top pair of wings of true bugs, which have their basal ends thickened and their distal ends membraneous.

Hemimetabola. Insects with an incomplete metamorphosis.

Herbaceous. Referring to plants with soft stems, such as annuals, biennials, and perennials that normally die back to the ground in the winter.

Herbicide. Any agent or chemical used in the destruction or control of weeds.

Hermaphroditism. Male and female reproduction organs in the same individual.

Heteroecious. Requiring two or more unrelated hosts for completing the life cycle.

Hibernation. The period of suspended animation in insects and in some higher animals in response to seasonal low temperatures.

Holocarpic reproduction. A process in fungi in which the entire thallus segments into spores.

Holometabola. Insects with a complete metamorphosis.

Homologues. A series of organic compounds with similar properties but differing from each other by some radical.

Honeydew. A liquid excretion from the anus of many Homoptera.

Host. Any plant or animal attacked by a parasite or a predator.

Hyperplastic. Refers to type of symptom characterized by the abnormal multiplication of cells.

Hypha (pl. *hyphae*). A single thread of fungous mycelium.

Hypoplastic. Refers to type of symptom characterized by the underdevelopment of cells, tissues, or organs.

Hypostome. A pronglike structure bearing recurved teeth forming part of the mouthparts of Ixodid ticks.

Immune. Not subject to attack or infection by a given organism.

Imperfect stage. The period in the life cycle of a fungus during which spores are produced asexually.

Incomplete metamorphosis. Similar to gradual metamorphosis but generally used to denote the growth changes of naiads.

Incubation stage or period. The period beginning when inoculation occurs and ending when symptoms of disease can first be observed. In birds, the period of embryonic development within the egg.

Infection court. Any place where an infection may take place.

Infection stage. The period during which the disease-producing organism causes symptoms of disease to appear in a plant.

Infective juvenile. The immature stage of a nematode capable of invading host tissue.

Inoculation stage. The period during which inoculum is being transferred from its source to the infection court.

Inoculum. Spores or any part of a pathogene that can infect plants.

Insectivorous. Feeding on insects.

Instar. The stage of an insect between successive molts.

Isthmus. The portion of the esophagus in nematodes between the metacorpus and basal bulb.

Ixodid ticks. Ticks belonging to the family Ixodidae or hard ticks as differentiated from the ticks belonging to the family Argasidae or soft ticks.

Karyogamy. The fusion of the two haploid nuclei of a dikaryon in a fungus to form a diploid spore.

Koch's postulates. Rules to be followed to prove the pathogenicity of a microorganism.

Larva (pl. *larvae*). The growing, worm-like stage of insects with a complete metamorphosis. Also, the newly hatched six-legged stage of mites and ticks.

LC_{50}. Lethal concentration of a toxicant in air or liquid that will kill 50% of the test organisms by inhalation or otherwise.

LD_{50}. Lethal dose of a toxicant that will kill, orally or dermally, 50% of the test organisms.

Lesion. Any break in the epidermis of a plant; a localized, diseased area.

Lethal. Deadly.

Life cycle, life history. The complete succession of events in the life of an organism.

Lumen. The fine tubular passage in the stylet and esophagus of nematodes that conducts nutrients.

Maggot. The growing stage or larva of a fly.

Mandibles. The heavy pair of biting or chewing organs in an insect's mouth.

Medium bulb. The metacorpus of a nematode.

Mesothorax. The middle section of the thorax of an insect that bears the top wings and the middle pair of legs.

Metacorpus. The posterior part of the corpus in nematodes usually ovate in shape and frequently containing a valve-like structure.

Metamorphosis. Any conspicuous changes in form or structure during the growth of animals.

Metathorax. The last division of the thorax of insects, bearing the second pair of wings and the third pair of legs.

Micron. A unit of measurement equal to one one-thousandth of a millimeter.

Mildew. A fungous disease characterized by the appearance of a white, mycelial growth and spores on the surface of infected plant parts.

Mold. A fungus with conspicuous mycelium or spore masses.

Molt. The shedding of the skin of insects and snakes in the process of growth, or of the feathers of birds and the fur of mammals.

Monodelphic. With one ovary, in nematodes.

Monoecious. Refers to plants that have separate staminate and pistillate flowers on the same plant; also to rusts that have all stages of their life cycle on a single species of plant.

Mosaic disease. One characterized by variegated patterns of green and yellow on the foliage of plants affected by certain viruses.

Mycelium. A mass of hyphae that forms the body of a fungus.

Mycoplasmalike organisms. Organisms that resemble both bacteria and viruses in some of their characteristics.

Naiad. The aquatic growing stage of insects with an incomplete metamorphosis.

Necrotic. Having symptoms characterized by the death or disintegration of cells or tissues.

Nematode worms, nemas. Microscopic roundworms, many of which attack plants.

Nestling birds. Newly hatched birds that are confined to their nests.

Nodus. A heavy cross vein near the middle of the upper margin of the wings of Odonota.

Non-infectious disease. A disease that cannot be transmitted from one plant to another; a disease caused by a physiopath.

Non-septate. Without cross-walls.

Nymph. The growing stage of insects with a gradual metamorphosis.

Ocelli. Small, single visual units located on the heads of spiders and many insects.

Oesophagus. The part of the alimentary canal that connects the mouth with the intestinal tract.

Oöspore. A resting spore produced by sexual reproduction in the downy mildews and related fungi.

Ovicide. Any factor or chemical that destroys eggs.

Oviparous. Reproducing from eggs laid by an animal.

Ovipositor. An organ, present in many insects, specialized for depositing eggs in plant structures or in the ground.

Ovoviviparous. Reproducing living young from eggs that hatch within the female's body.

Parasite. Any organism that lives on or in the body of another organism and obtains nourishment from it.

Parasitoid. An organism that feeds in or on another organism eventually destroying it.

Pathenogenesis. The reproduction of young from unfertilized eggs.

Pathogene. Any organism that is capable of causing disease.

Pathogenesis. The period during which the pathogene is actively attacking the living tissues of its host.

Paurometabola. Insects with a gradual metamorphosis.

Perfect stage. The part of the life cycle of a fungus during which spores are produced sexually.

Perithecium (pl. *perithecia*). The flask-like, fruiting body of a fungus, containing asci, usually the overwintering stage of certain fungi such as the one causing apple scab.

Pest. Any organism injuring animals, plants or plant products.

Phasmids. A pair of tiny lateral, rounded structures near the tail-end of nematodes believed to be sensory in nature.

Pheromone. A hormonal substance secreted by insects used in attracting the opposite sex or in eliciting other responses.

Physiogenic disease. A disease produced by a physical or an environmental factor.

Physiopath. Any physical or environmental factor capable of producing symptoms of disease in plants.

Phytotoxicity. Toxicity to plants.

Plasmogamy. The initiation of the diploid phase in the life cycle of fungi, the process in which two haploid nuclei of two different mating types come together in one cell but divide separately as dikaryons until they fuse in Karyogamy.

Plasmodium (pl. *plasmodia*). The naked amoeboid thallus of slime molds and related forms.

Poison bait. A substance that has a poison mixed with it that is attractive as a food for certain animals.

Predator. An animal (insect included) that attacks and kills another animal in order to obtain food.

Procorpus. The portion of the corpus of the esophagus in nematodes located between the base of the stoma and the metacorpus.

Prolegs. The fleshy, unsegmented leg-like structures on the abdomens of some larvae.

Pronotum. The dorsal sclerite of the prothorax.

Prothorax. The first segment of the thorax, which bears the first pair of legs.

Pupa (pl. *pupae*). The stage during which an insect with complete metamorphosis is transforming from the larval to the adult stage.

Puparium (pl. *puparia*). The darkened and hardened last larval skin of a maggot, which protects the pupa of true flies.

Pycnidium (pl. *pycnidia*). Flask-like fruiting body containing conidia.

Pyriform. Pear-shaped.

Race or strain. Organisms of the same species and variety that differ in their ability to produce disease in varieties of a given host or that differ in their reaction to insecticides or fungicides.

Residual contact pesticide. A pesticide with extended residual toxicity through tarsal or body contact.

Resistant. The ability of a host to suppress or retard the injurious effects of an organism.

Restricted-use pesticides. Pesticides designated by the EPA or a similar state agency, that can be applied only by certified applicators for specific purposes because of their inherent toxicity or potential hazard to the environment.

Rhizome. An underground stem that sends up new shoots.

Rickettsiae. Organisms with scalloped cell walls that resemble bacteria.

Rodenticides. Chemicals used to destroy rodents and related animals.

Roguing. The removal of undesired individual plants from a group of plants.

Russetting. Brownish roughening on the skins of fruits as a result of disease or injury.

Saprogenesis. The period in the life cycle of a pathogene when it is not associated with living tissues of its host. This may include a period of dormancy.

Saprogenic (Saprophagous). Obtaining nourishment from nonliving organic matter.

Sclerite. A hardened body wall plate bounded by sutures.

Sclerotium (pl. *sclerotia*). A hard compact mass of fungal tissue capable of surviving adverse conditions.

Sclerotized. Hardened dense condition of cuticular structures.

Scutellum. A more or less triangular dorsal sclerite between the base of the wings of Hemiptera, Homoptera, Coleoptera and some Diptera.

Sedentary. Term referring to insects or nematodes that do not change their feeding site after they have become established.

Segmented. Divided into distinct divisions.

Selective herbicide. One that destroys certain species of plants or weeds and leaves others relatively unharmed.

Septate. Possessing cross-walls.

Seta (pl. *setae*). Hairlike bristles on the cuticle of insects.

Sign. Evidence of disease as indicated by the presence of the disease-producing organisms or of any of their parts or products.

Slime flux. A fermenting exudate from tree trunks.

Slurry. A thick suspension of a finely divided substance in a liquid.

Sorus (pl. *sori*). Fruiting bodies of rust and smut fungi in which masses of rusty or black spores are produced.

Species. One kind of plant or animal. Abbreviated as sp. singular and spp. plural. Spp. following genus name means that a number of species of that genus are indicated.

Spicules. A pair of forcep-like claspers near anus of male nematodes.

Spinneret. Organs found in certain insect larvae and in spiders, which are used in spinning silken threads.

Spiracles. The external openings to the breathing organs of insects and related forms.

Spore. A single-to-many-celled body capable of reproducing the fungus or other lower plant from which it originated.

Sporophyte. The sexual or perfect phase in the life cycle of a fungus.

Sterilant. Any agent or chemical that destroys all living organisms in a substance or renders it barren.

Sternum (pl. *sterna*). The ventral body sclerite of an insect.

Stigma. A darkened area in the upper margin near the end of the wings of Odonata and some Hymenoptera.

Stigmata. Breathing pores located on the dorsal side of arachnids near the mouth parts.

Stolon. A creeping stem that sends down roots and produces new shoots.

Stomach poison. A poison that takes effect when swallowed.

Stoma. The mouth or buccal cavity of a nematode.

Stomata (sing. *stoma*). Minute pores present in the leaves of plants, utilized in the exchange of gases for respiration and photosynthesis.

Strain, see Race.

Stroma (pl. *stromata*). A mass of dark fungal tissue in which fruiting bodies are produced.

Stylets. The slender, hollow, piercing and sucking organs possessed by insects and nematodes that feed on the sap of plants.

Subspecies. Groups in an animal species that differ from each other in certain minor details usually because of geographical location.

Suscept. A more precise term for host—any plant liable to infection by a given organism.

Symptoms. Evidence of disease in plants expressed by the reaction of the plants to the presence of the irritating factor or organism.

Synergist. Any substance that increases the toxic effects of a pest-control chemical.

Synthetic. Compounded in the laboratory, as opposed to occurring naturally.

Systemic disease. One in which the pathogene spreads throughout the plant body.

Systemic pesticide. One that can be absorbed and translocated within plants or animals to destroy pests feeding on them.

Tarsus (pl. *tarsi*). The jointed "foot" that bears the claws of an insect.

Tergum (pl. *terga*). The dorsal body sclerite of an insect.

Thallus (pl. *thalli*). The body of a fungus.

Thorax. That portion of an insect's body which lies between the head and abdomen and bears the legs and wings.

Tibia (pl. *tibiae*). The long, slim segment of an insect's leg to which the tarsus is attached.

Tolerance. Amount of toxic residue allowable on or in edible substances under the law.

Toxic. Poisonous.

Toxin. Poison produced by an organism.

Tracheae. The larger respiratory tubes leading from the spiracles.

Trade name. Name given to a product sold by a company to distinguish it from similar products made by other companies.

Unsulfonated residue. A measure of the saturated hydrocarbons in an oil that does not react or combine with sulfuric acid.

V. The symbol for the vulva in nematodes with the location expressed as a percentage of the total body length as measured from the anterior end.

Varieties. Plants in a species that differ from each other in certain details such as form, color, fruit, size, fruit flavor, etc.

Vascular. Refers to conducting tissues (phloem, xylem) of plants.

Vector. Any carrier of a disease-producing organism.

Ventral. Referring to the underside of an organism.

Viability. State of being alive.

Virions. The particles of a virus.

Viroids. A class of subviral plant pathogens composed of naked RNA.

Virulent. Strong ability to produce disease.

Viviparous. Reproducing living young that undergo embryonic development in the abdominal cavity of the female.

Wettable powder. One that is easily wetted by water and will go into suspension.

Zoöspore. Spore capable of independent movement in water; a swarmspore.

Zygote. The sexual (2n) stage of a fungus formed by the union of two gametes.

Useful Conversions and Measurements

Metric Equivalents

1 pound per acre	=	1.120 kilograms per hectare
1 pound per 1000 square feet	=	48.8 kilograms per hectare
1 ounce per 100 square feet	=	30.5 kilograms per hectare
1 kilogram per hectare	=	0.8922 pounds per acre
100 gallons per acre	=	935 liters per hectare
1 hectare	=	2.471 acres
1 kilogram	=	2.2046 pounds
1 pound	=	0.4536 kilograms
1 acre	=	0.4047 hectare
1 gallon	=	3.785 liters

Miscellaneous Measures

1 acre = 43,560 square feet = 4,840 square yards = 160 square rods
1 tablespoonful = 3 teaspoonfuls
1 fluid ounce = 2 tablespoonfuls
1 cupful = 8 fluid ounces = 16 tablespoonfuls
1 pint = 2 cupfuls = 16 fluid ounces
1 U.S. gallon = 231 cubic inches = 8.34 lb of water
1 imperial gallon = 277.4 cubic inches = 10.0 lb of water
Degree C = 5/9X (Degree F -32)
Degree F = 9/5X (Degree C) $+32$

Metric Length

1 inch	=	2.54 centimeters
1 foot	=	.305 meter
1 yard	=	.914 meter
1 mile	=	1.609 kilometers
1 centimeter	=	.394 inch
1 meter	=	3.281 feet
1 meter	=	1.094 yards
1 kilometer	=	.621 mile

1 mile = 5,280 feet = 1,760 yds = 320 rods
1 rod = 16½ feet = 5.0325 meters

Measure of Surface (area)

144 square inches	=	1 square foot
9 square feet	=	1 square yard
30¼ square yards	=	1 square rod
160 square rods	=	1 acre
43,560 square feet	=	1 acre
640 square acres	=	1 square mile

TABLE C.1

CONVERSION TABLE, LIQUID MEASURE

	Ml	Tablespoon	Ounce	Cup	Pint	Quart	Liter	Gallon
Milliliter	1	0.067	0.033	0.0042	0.0021	0.0011	0.0010	0.0002
Tablespoon	15	1	0.500	0.0625	0.0312	0.0156	0.0149	0.0039
Ounce	29.57	2	1	0.1250	0.0625	0.0312	0.0295	0.0078
Cup	236.5	16	8	1	0.5000	0.250	0.2364	0.0625
Pint	473	32	16	2	1	0.500	0.474	0.1250
Quart	946	64	32	4	2	1	0.943	0.2500
Liter	1000	66.670	33.82	4.2300	2.11	1.060	1	0.2640
Gallon	3785	256	128	16	8	4	3.785	1

Source: Agric. Extension Service, Univ. of Wyoming

TABLE C.2

CONVERSION TABLE, DRY MEASURE

	Gram	Ounce	Pound	Kilogram
Gram	1	0.0352	0.0022	0.0010
Ounce	28.35	1	0.0625	0.2835
Pound	453.6	16	1	0.4536
Kilogram	1000	35.27	2.20	1

Selected References

AGRIOS, G.N. 1978. Plant Pathology, 2nd Edition. Academic Press, New York.

ANDERSON, J. F., and KAYA, H. K. (Editors). 1976. Perspectives in Forest Entomology. Academic Press, New York.

ANON. 1968. Control of Plant Parasitic Nematodes, Vol. 4. National Academy of Science, Washington, D.C.

ANON. 1974. Sterility Principle for Insect Control. Unipublishers, New York.

ANON. 1975. Forest Pest Control, Vol. 4. National Academy of Science, Washington, D. C.

ANON. 1976. Controlling Fruit Flies by the Sterile Male Technique. Unipublishers, New York.

BAKER, K. F. 1974. Biological Control of Plant Pathogens. Freeman Press, New York.

BAKER, K. F. et al. (Editors). 1976. Annu. Rev. Phytopath. 14, Annual Reviews, Palo Alto, California.

BARBOSA, P., and PETERS, T. M. 1972. Readings in Entomology. W. B. Saunders Co., Philadelphia.

BILLINGS, S.C. 1978. Pesticide Handbook—Entoma. Entomological Society of America, College Park, Md.

BLOWER, J. G. (Editor). 1972. Myriapoda. Academic Press, New York.

BORER, D. W., DELONG, D. M., and TRIPPLEHORN, C. A. 1976. An Introduction to the Study of Insects, 4th Edition. Holt, Rinehart and Winston, New York.

BORROR, D. J., and WHITE, R. E. 1970. A Field Guide to the Insects. Houghton Mifflin Co., Boston.

BOWERS, W. S., et al. 1976. Discovery of anti-juvenile hormones in plants. Science 193, No. 4253, 542–547.

CARTER, W. 1973. Insects in Relation to Plant Disease. Wiley-Interscience, New York.

CHU, H. F. 1949. How to Know the Immature Insects. Wm. C. Brown Co., Dubuque, Iowa.

CHUPP, C., and SHERF, A. F. 1960. Vegetable Diseases and Their Control. Ronald Press, New York.

CLARK, L. R., et al. 1974. The Ecology of Insect Populations in Theory and Practice. Halstead Press, New York.

CROLL, N. A. (Editor). 1976. The Organization of Nematodes. Academic Press, New York.

DAVIDSON, G. 1974. Genetic Control of Insect Pests. Academic Press, New York.

DAVIDSON, R. H., and PEAIRS, L. M. 1966. Insect Pests of Farms, Garden and Orchard, 6th Edition. John Wiley and Sons, New York.

DEONG, E. R., et al. 1972. Insect, Disease and Weed Control, 2nd Edition. Chemical Publications Co., New York.

DETHIER, V. G. 1976. Man's Plague? Insects and Agriculture. Darwin Press, Princeton, N.J.

DICKSON, J. G. 1952. Diseases of Field Crops, 2nd Edition. McGraw-Hill, New York.

EBELING, W. 1975. Urban Entomology. Div. of Agric. Sci. Cal., Univ. of Cal.

EDWARDS, C.E. 1974. Persistent Pesticides in the Environment. CRC Press, Cleveland, Ohio.

ELZINGA, R.J. 1978. Fundamentals of Entomology. Prentice-Hall, Englewood Cliffs, New Jersey.

FUNDER, S. 1968. Practical Mycology, 3rd Edition. Hafner Publishing Co., New York.

GLASS, E. W. (Editor). 1975. Integrated Pest Management: Rationale, Potential Needs and Implementation. Entomological Soc. of America, College Park, Md.

HICKIN, N.E. 1974. Household Insect Pests. St. Martin's Press, New York.

HILL, D. S., 1975. Agricultural Insect Pests of the Tropics and Their Control. Cambridge University Press, Cambridge, England.

HORN, D. J. 1976. Biology of Insects. Saunders, New York.

HUFFAKER, C. B., et al. 1976. Theory and Practice of Biological Control. Academic Press, New York.

JACOBSEN, J. S., and HILL, C. C. 1970. Recognition of Air Pollution Injury to Vegetation: A Pictorial Atlas. Air Pollution Control Assoc., Pittsburgh.

JAMES, M. T., and HARWOOD, R. F. 1969. Medical Entomology. Macmillan Co., New York.

JENKINS, W. R., and Taylor, D. P. 1967. Plant Nematology. Reinhold Publishing Co., New York.

JEPPSON, L. R., KIEFER, H. H., and BAKER, E. W. 1975. Mites Injurious to Economic Plants. University of Cal. Press, Berkeley, Cal.

JOHNSON, W. T., and LYON, H. H. 1976. Insects That Feed on Trees and Shrubs. Cornell University Press, Ithaca, N.Y.

JONES, F. G. W., et al. 1974. Pests of Field Crops, 2nd Edition. St. Martin's Press, New York.

KLOTZ, L. J. 1961. Color Handbook of Citrus Diseases, 3rd Edition. University of California, Berkeley, Cal.

LITTLE, V. A. 1972. General and Applied Entomology, 3rd Edition. Harper and Row Publishers, New York.

MARTIN, H. (Editor). 1972. Insecticide and Fungicide Handbook, 4th Edition. Lippincott, New York.

MAI, W. F. 1975. Pictorial Key to Genera of Plant Parasitic Nematodes, 4th Rev. Edition. Cornell University Press, Ithaca, N.Y.

MATSUMURA, F. 1975. Toxicology of Insecticides. Plenum Publishers, New York.

MATTHEWS, R. E. 1970. Plant Virology. Academic Press, New York.

METCALF, C. L., et al. 1962. Destructive and Useful Insects, 4th Edition. McGraw-Hill, New York.

METCALF, R. L., and LUCKMANN, W. H. (Editors). 1975. Introduction to Insect Pest Management. John Wiley and Sons, New York.

METCALF, R. L., and McKELVEY. J. J. (Editors). 1976. The Future for Insecticides: Needs and Prospects. Wiley-Interscience, New York.

MOORE-LANDECKER, E. 1972. Fundamentals of the Fungi. Prentice-Hall Inc., Englewood Cliffs, N. J.

MUNRO, J. W. 1973. Pests of Stored Products. St. Martin's Press, New York.

PAL, R. and WHITTEN, M. J. 1974. The Use of Genetics in Insect Control. North-Holland Publications Co., New York.

PFADT, R.E. (Editor). 1978. Fundamentals of Applied Entomology, 3rd Edition. Macmillan Co., New York.

PIMENTEL, D. (Editor). 1975. Insects, Science and Society: Proceedings. Academic Press, New York.

PIMENTEL, D. 1976. World food crisis: energy and pests. Bul. Entomol. Soc. Am. 22, No. 1, 20–26.

PIRONE, P.P. 1978. Diseases and Pests of Ornamental Plants, 5th Edition. Wiley-Interscience, Somerset, New Jersey.

PRICE, P. W. 1975. Insect Ecology. Wiley-Interscience, New York.

PYENSON, L. L. and Barke, H. E. 1976. Laboratory Manual for Entomology and Plant Pathology. AVI Publishing Co., Westport, Conn.

ROBERTS, D.A. 1978. Fundamentals of Plant Pest Control. W.H. Freeman, San Francisco.

ROBERTS, D.A., and BOOTHROYD, C.W. 1972. Fundamentals of Plant Pathology. W.H. Freeman and Co., San Francisco.

BERG, G.L. (Editor). 1979. Farm Chemicals—Dictionary of Pesticides. Meister Publishing Co., Willoughby, Ohio.

SHOREY, H. H., and McKELVEY, J. J. 1976. Chemical Control of Insect Behavior: Theory and Application. Wiley-Interscience, New York.

SMITH, R. F., MITTLER, T. E., and SMITH, C. N. (Editors). 1975. Annual Review of Entomology, Vol. 20, Annual Reviews, Palo Alto, California.

SPRAGUE, H. B. (Editor). 1964. Hunger Signs in Plants, 3rd Edition. David Mckay Co., New York.

STREETS, R. B. 1972. The Diagnosis of Plant Diseases. The University of Arizona Press, Tucson, Ariz.

STROBEL, G. A., and MATHRE, D. E. 1970. Outlines of Plant Pathology. Van Nostrand Reinhold Co., New York.

SWEETMAN, H. L. 1965. Recognition of Structural Pests and Their Damage. Wm. C. Brown Co. Dubuque, Iowa.

TATTAR, T.A. 1978. Diseases of Shade Trees. Academic Press, New York.

THORNE, G., 1961. Principles of Nematology. McGraw-Hill, New York.
TUITE, J. 1969. Plant Pathological Methods. Burgess Publishing Co., Minneapolis, Minn.
VAN DER PLANK, J. E. 1975. Principles of Plant Infection. Academic Press, New York.
WALKER, J. C. 1969. Plant Pathology, 3rd Edition. McGraw-Hill Book Co., New York.
WALLACE, H. R. 1974. Nematode Ecology and Plant Disease. Crane-Russak Co., New York.
WARE, G. W. 1975. Pesticides, An Auto-Tutorial Approach. W. H. Freeman and Co., San Francisco.
WATSON, T. F., MOORE, L., and WARE, G. W. 1976. Practical Insect Pest Management. W. H. Freeman and Co., San Francisco.
WAYS, M. J. 1976. Entomology and the world food situation. Bull. Entomol. Soc. Am. 22, No. 2, 125–129.
WELLMAN, F. L. 1972. Tropical American Plant Disease. Scarecrow Publishing Co., Metuchen, N.J.
WESTCOTT, C. 1973. The Gardener's Bug Book. Doubleday, Garden City, N.Y.
WESTCOTT, C., 1971. Plant Disease Handbook, 3rd Edition. D. Van Nostrand Co., New York.
WOODS, A. 1974. Pest Control: A Survey. Halsted Press, New York.

Professional Journals

Environmental Entomology. Entomological Society of America, College Park, Md.
International Journal of Acarology. Acarology Society of America, Oak Park, Mich.
International Journal of Nematology. The American Phytopathological Society, St. Paul, Minn.
Journal of Economic Entomology. Entomological Society of America, College Park, Md.
Phytopathology. The American Phytopathological Society, St. Paul, Minn.

Index

Other AVI Books